# TEFL Tourism

# TEFL Tourism

*Principles, Commodification and the Sustainability of Teaching English as a Foreign Language*

**Hayley Stainton**

CABI

**CABI is a trading name of CAB International**

CABI
Nosworthy Way
Wallingford
Oxfordshire OX10 8DE
UK

Tel: +44 (0)1491 832111
Fax: +44 (0)1491 833508
E-mail: info@cabi.org
Website: www.cabi.org

CABI
745 Atlantic Avenue
8th Floor
Boston, MA 02111
USA

Tel: +1 (617)682-9015
E-mail: cabi-nao@cabi.org



A catalogue record for this book is available from the British Library, London, UK.

ISBN-13: 9781786393227 (Hardback)
9781786393234 (ePDF)
9781786393241 (ePub)

Commissioning Editor: Claire Parfitt
Editorial Assistant: Tabitha Jay
Production Editor: James Bishop

Typeset by SPi, Pondicherry, India
Printed and bound in the UK by Severn, Gloucester.

# Contents

# 1

# Introduction

The burgeoning globalisation of the world economy has seen societal English-speaking capabilities increase in importance throughout many countries. Similar to other associated industries such as volunteer, responsible or sustainable tourism, the focus of teaching English as a foreign language (TEFL) has in many cases shifted away from its traditional altruistic and philanthropic perspective, where education was the priority, to a postmodern commercialised commodity, acting as a resource that can be produced, controlled, distributed and valued.

Essentially a combination of the notions of tourism and English language teaching, TEFL tourism is introduced in this book as a micro-niche tourism form designed to satisfy the demands for English language teachers around the globe. Through demonstrating parallels between TEFL tourism and other alternative tourism types, such as volunteer tourism, a number of cautionary approaches are advised in order to facilitate appropriate sustainable management of the TEFL tourism industry.

## TEFL Tourism

The limited academic reference to TEFL and its links with tourism to date is surprising, given the extent and importance of the English language teaching industry worldwide and taking into account the size of the tourism industry nowadays – arguably the biggest service industry worldwide. While there is ample evidence of the amalgamation of the two concepts of TEFL and tourism, as demonstrated throughout the chapters of this book, the majority of existing research revolves around pedagogical practice or development agendas. It is this gap in the literature which inspired the production of this book, with the ambition of raising awareness in both academic and non-academic communities of the benefits and pitfalls that the tourism element may present.

The English language has become an integral part of world society, with many destinations putting significant focus on the expansion of this as part of their development agendas. Enhanced English language capability opens up an array of doors for a destination in terms of economic and social development. This book outlines the role that TEFL tourism plays in modern-day capitalist society. It is estimated that by 2020 there will be over 2 billion English language learners (British Council, 2014), and with this brings insatiable demand for teachers. This presents one of the most significant challenges to educators and citizens across the globe and many destinations are now looking towards tourists to fill employment gaps.

Rooted in the concepts of tourism and TEFL, TEFL tourism is one of many emerging niche tourism forms available to tourists. TEFL tourists vary considerably, as do all tourist types, and based on previous research this book delves into the typical expectations, motivations and typological features of TEFL tourists. This is, of course, subject to geographical and cultural adaptations, which add a layer of complexity to understanding the TEFL tourism industry in its entirety. However, by providing a foundation for the advanced comprehension of the TEFL tourism industry, as is presented throughout this book, contextualisation can be applied.

In theory, the utilisation of tourists to meet the demands for English language teachers appears to provide an excellent solution to a problem. However, there are many lessons to be learned from the wider tourism and education industries. It is not suggested that TEFL tourism is a bad thing, far from it. This book simply airs caution emphasising that if managed carefully, it may yield the optimum outcomes intended. Sustainable management is paramount.

There is a distinct paradox between the benevolent intentions of philanthropic tourists who wish to do good and to help an otherwise struggling community, and those of profit-oriented organisations. Converging the two into one setting raises questions regarding how this may work in practice, and whether indeed the positive impacts may well be diluted by the very presence of capitalism. There is certainly an argument to say that this is the way that modern-day society operates and that it is an inevitable developmental progression. But there is also ample evidence of where such market relationships have failed to be successful.

Although there is little research to draw upon with regard to TEFL as a form of tourism specifically, it does have allies within the tourism industry. This book draws on many examples taken from the volunteer tourism industry, which has been subjected to its fair share of criticisms in recent years. While TEFL tourism may sometimes fall under the umbrella of volunteer tourism, when there is no financial benefit to the tourist, it can also be a very different entity. As such, literature from associated industries, such as volunteer tourism, is referred to in an advisory manner only, emphasising the need for concepts to be applied to, and assessed in relation to, TEFL tourism.

One way to attempt to further understand the nature of the industry is to analyse the key stakeholders. Educational stakeholders are examined, including the host destination's government which may prioritise TEFL tourism as part of its development agenda, and stakeholders in the tourism-generating region. It is clear that the nature of the TEFL experience is not clearly understood by university staff and teacher training institutions, and aspects that may be core parts of the curriculum such as differentiation or Office for Standards in Education, Children's Services and Skills (Ofsted) requirements may not be a part of a TEFL tourist's role. This book aims to provide some clarification to education-based stakeholders. Another key stakeholder is the TEFL agency. Demonstrating many parallels with the traditional package holiday model, this book assesses the commodification of TEFL teaching, discussing the role played by businesses who facilitate this. The book thus aims to provide guidance for said organisations, encouraging sustainable practices and management. Lastly, central to the TEFL tourism experience are the tourists themselves, along with their students. The latter part of the book examines the TEFL tourist, presenting a relevant typology that can be used to enhance understanding of who the tourists are and their motivations and experiences. This information can then be useful to the stakeholders just discussed in order to more effectively manage their operations.

## Historical Context of Niche Tourism

The state of the travel and tourism industry has altered considerably over recent years due to changes in preferences and demand by the consumer, the way that technology is utilised and lifestyle choices. This new tourism can be seen within the emergence of niche tourism. Niche tourism as a disparate concept and theory was not recognised in full until the early 2000s. It was in 2002 that the first niche tourism conference was held (Crichton Tourism Research Centre at the University of Glasgow). This conference aimed to draw together for the first time academics from different tourism-related fields to discuss this new type of tourism, thus enabling some of the first collective thinking on the topic.

Novelli's (2005) preliminary text on addressing the concept of niche tourism provided the foundations for the analysis of the niche tourism industry through a series of theoretically underpinned case studies. She suggests that the term 'niche tourism' has its roots in the concept of 'niche marketing' – it is effectively a counterpart to the undifferentiated mass tourism product and is designed to meet the needs of individual market segments or niches. The niche tourism spectrum, or continuum, is made up from large market sectors or macro-niches (i.e. cultural tourism), which can then be subdivided into micro-niches (e.g. TEFL tourism).

The concept of niche tourism derives originally from marketing. Hutchinson (1957) first introduced the 'niche' concept, referring to an optimum location that an organism can exploit against its competitors. Keegan et al. (1992) later described a 'niche' as a small market not currently served by competing products. The term 'niche' then began to appear in business literature, which is from where the phrase 'niche marketing' was derived. Most research published since the 1990s has emphasised new segmentation within existing marketplaces rather than identifying entirely new markets, as this book does with TEFL tourism. There are generally four key factors attributed to market segmentation: socioeconomic/demographic; geographic; psychographic and psychological and behavioural.

Niche marketing has continued to evolve since the 1980s and organisations have come to realise that they need to tailor their products and services to ever-changing customer demands and tastes. In fact, many companies have exchanged traditional mass marketing techniques for niche marketing in a range of contexts. Niche marketing (and indeed niche tourism) plays a key role in a product's lifecycle. Many companies will have niche markets at the initial stage of their product before that product develops into maturity and then saturation starts, when successful companies need to reinvent their products, often returning to the niche market.

The application of these niche markets to tourism introduces new tourist typologies, whereby tourism clusters or groups are characterised by aspects such as behaviours or motivations. According to Read (1980), special interest tourism was the predecessor of 'niche tourism'. Special interest tourism marked the transition from tourism as a commodified, mainstream offering to one that was specialised and unique. The rise of the mass tourism industry, largely fuelled by the growth in aviation, also gave rise to the backpacker tourist, perhaps one of the first special interest tourist types. During this time mass tourism began to be no longer seen as the dominant tourism form, and special interest tourism became the new tourism form of the 1980s and 1990s.

This trend has very much continued to this day. That is not to say that the mass tourism industry has dwindled, it is simply that consumers tend to prefer experiences that are better tailored to their individual needs and desires. Nowadays, niche tourism forms are more commonly referred to as 'alternative tourism', which attempts to balance mass tourism by seeking smaller numbers with higher spending power. Academic literature, however, has paid little attention to the positioning of niche tourism products. Tourist behaviour has produced a more segmented and sophisticated consumer market, which niche tourism has responded to (Sharpley and Telfer, 2002). Establishing defined and specific niches has allowed smaller, independent tourism-based organisations to feature in the competitive tourism marketplace. The development of niche tourism products is commonly associated with high levels of entrepreneurship and an individual desire to explore specific market niches (Novelli, 2005). There has been little research to date, however, examining the relationship between niche tourism development and entrepreneurship, an area that certainly warrants investigation within TEFL tourism.

## The Postmodern Tourist

Tourism emerged in response to the processes of industrialisation, urbanisation, and technical and technological developments of society during the Industrial Revolution and has continued to evolve in correlation with consumer demands and preferences throughout the years. Through the democratisation of the tourism and leisure industries, the tourism industry has become an increasingly important part of the worldwide economy, with serious social and political impacts. Contemporary tourism includes a wide variety of tourism forms and elements including means of transport, destinations and activities, which, depending on the context of time, gain or lose popularity.

Throughout the years we have continued to see changes in the economic, political, cultural and social landscapes of modernity. New cultures have emerged, characterised by a decreasing use of mass production, mass consumers, large towns and cities and material and technical advancement. There has also been an overall reduction in the likes of standardised production, urbanisation, industrialisation, centralisation and bureaucratisation. Instead these are increasingly being replaced with notions of flexibility, diversity, increased mobility, enhanced communication processes, decentralisation and internationalisation. This has resulted in new tourism forms, such as TEFL tourism, emerging. These tourism forms have the potential to replace or instigate change in existing tourism forms and fundamental tourism structures.

In contrast to Fordism, the system formulated in Henry Ford's automotive factories, whereby workers operate on a production line, performing specialised tasks repetitively, tourists nowadays demonstrate a post-Fordist attitude. This has been the dominant system of economic production, consumption and associated socio-economic phenomena in most industrialised countries since the late 20th century. Like the growth in the demand for special interest or 'niche' tourism, post-Fordism has brought with it new ways of looking at consumption and production. Rather than the tourism industry being viewed as a mass market to be served by mass production, tourists have begun to be viewed as different groups pursuing different goals who could be better served by small batches of specialised services. The popularity of the mass markets has become less important, while market tourism forms that bring with them alternative benefits such as CV building or a feeling of being able to 'do good' are increasing in popularity. Tourists are more and more seeking experiences with greater value than is typically obtained via the traditional package holiday route.

The idea of the 'post-tourist' has developed in response to consumer attitudes and preferences in the postmodern era. Post-tourists are familiar with new technology and are particularly responsive to social media. In fact, the post-tourist does not even need to leave their home in order to experience the traditional objects of the tourist gaze. Increasingly, simulated tourist experiences are brought into homes through television, websites, online reviews, bloggers and influencers. Some argue that postmodern tourists are indeed the antithesis of mass tourists, demanding authentic experiences by venturing away from mass tourist areas and facilities.

Due to changes in technology, the postmodern tourist is far more aware of the options available to them than their modern counterparts. Consumers nowadays are frequently influenced by those they follow on social and traditional media channels. They have a wealth of information accessible to them at the click of a button via the internet and their powers as a consumer are arguably stronger than ever before. There has also been a significant shift in societal attitudes, with people becoming more conscious of their actions and behaviours. All of this has led to the growth of niche tourism forms, in particular those with a developmental or sustainability focus. In order to be successful, tourism operators and stakeholders must respond to the demands of the present-day consumer, and this book intends to contribute to this through an exploration of the TEFL tourism industry.

## Operators, Communities and TEFL Tourism

While in essence the product of English language is intangible, aligned with the ideals of enhanced social development and business prospects, there is a strong body of evidence demonstrating that in many instances it has now been commodified into a product for sale in an ever-competitive capitalist market. In many contexts, language is now treated as an economic resource, cultivated for material profit, or acquired as a skill to be offered on the market (Park and Wee, 2012). In essence, the whole TEFL process is organised and reshaped in order to conform to commercial logic. It has become a business. The growth of the capitalist mode of production has expanded to such an extent that it has now begun to infiltrate markets previously naïve to commercial exploits. As a result, the social organisation of TEFL has been completely refashioned and reshaped in a commercial (profit- and market-oriented) way; the 'old ways of producing' have now been completely displaced by this new industrialism.

Commodifying TEFL in the context of tourism has enabled it to become an output of production. This is represented through the process of combining various material inputs (transport, accommodation, etc.) and immaterial inputs (plans, know-how) in order to make something for consumption (the output/TEFL product). Many organisations and individuals are now capitalising on the demand for English language by setting up new businesses, franchises or subsidiaries where teaching of English is sold on the respective market. The problem is that the motives of profit-focused organisations are not necessarily aligned with those of the community.

This book discusses at length the way in which the TEFL product has been commodified and the impacts that this may have. While some reference is placed at various times on the community, this is an area warranting additional research. Of course, there are concerns of increased crime, feelings of anger towards tourists, increased litter, cultural changes, etc. that may arise as a direct result of TEFL tourism, and this is widely recognised in the literature on the tourism industry in general. What is not as widely recognised, however, is the impact of differences between tourist and local salaries, of accelerated globalisation and of changes to infrastructure and customs as a result of TEFL tourism. This is not to mention the educational outcomes and how these might be affected by the presence of TEFL tourism. It is essential to take all stakeholders into consideration and to manage the industry as well as possible, based on existing and associated research and knowledge.

## A New Approach

The contribution to knowledge facilitated through the production of this book enables various TEFL stakeholders to have a more comprehensive understanding of the TEFL experience than literature has facilitated to date. It also provides an opportunity for academics and industry professionals alike to more accurately assess and manage the TEFL sector. Dominant stakeholders, who can be defined as any group or individual who can affect or are affected by the achievement of the organisation's objectives (Freeman, 1984, p. 46) include the TEFL teachers, TEFL recruitment agencies, future employers or education recruitment teams of the TEFL teacher and the Thai educational system and government.

The first of its kind, this book has been developed to encourage a new approach. It aims to encourage critical thinking, accurate assessment and sustainable management. While underpinned by academic research and literature, the intention is to provide more than discussion in the classroom: it is designed to facilitate real-world impact. It is eagerly anticipated that as a result of the publication of this seminal text on TEFL tourism, subsequent research can be undertaken by both scholars and industry practitioners. Said research can then be utilised to guide and inform the management of the TEFL industry and its associated communities.

Many lessons have been learned from a variety of tourism forms in the past falling within various parts of the mass-niche spectrum. TEFL tourism is a relatively new concept that is emerging at a rapid rate and it is imperative that stakeholders learn from these lessons of the past, applying them to this new tourism sector and thus ensuring the most optimum management approaches are adopted. As an introductory text, this book is not comprehensive, nor exhaustive, in its content. Rather, it brings about a new way of thinking in an industry that has lacked critical attention to date. As such, further research in the area of TEFL tourism is warmly welcomed and indeed necessary to ensure sustainability within the industry.

# 2

# The Global TEFL Tourism Industry

## Defining TEFL Tourism

The term TEFL tourism is by no means an elusive concept, yet it has to overcome the subjective limitations placed upon it as a result of the vast industry of English language teaching (ELT), with its encompassed geographical, cultural and linguistic differences. In fact, the very nature of ELT is continuously evolving as a result of scholarly research, development agendas and the increasing commercialisation of the sector. Susan Griffith, in her guide to teaching English abroad (Griffith, 2014), emphasises the confusing nature of definitions used in the ELT industry. Like Griffith, this book refers to the acronym TEFL to describe the act of a person who teaches English as a foreign language. However there are many acronyms used within the context of ELT (Table 2.1), and although alternatives may be adopted, the act of ELT can still qualify as TEFL tourism. The intention is that the term TEFL tourism encompasses all acts of teaching English as a foreign language which involve an element of tourism (the tourism element is discussed shortly).

The above acronyms are frequently used interchangeably, with little differentiation between the act of teaching English among them. For those who choose to travel abroad to teach, the most common title is TEFL, hence the introduction of the term 'TEFL tourism' in this book. 'TEFL' opportunities are promoted by a number of well-known organisations associated with travel and tourism, such as Real Gap, i-to-i and Gap 360. In the UK, however, the mainstream expression used is ELT as it is preferred by academic associations such as the University of Cambridge and publishers of journals in the field. TESL, TESOL and ESL tend to be associated with people who need to learn English in their everyday lives, and teachers often work with immigrants who have relocated to English-speaking countries or those who live in former colonies where English continues to be the medium of instruction in schools. ESP is largely focused on matching language needs with various professions such as business, banking, tourism, medicine, aviation, science, technology, etc. According to Griffith (2014), English for business is the most prominent in this sector and English for shopping, as commonly offered in Japan, is the least important. EAP is aimed towards students who wish to attend English-speaking universities in their home country or overseas and is often run by government-funded organisations such as the British Council.

While the concept of teaching can be deemed ambiguous in the global context, this is not the only barrier when defining TEFL tourism. Likewise, tourism has no universally accepted definition, owing to the complexity and individualism of the travellers themselves and the activities that they choose to undertake. The most widely utilised definition, proposed by the World Trade Organisation (WTO) and United Nations (UN) Statistics Division (1994), prescribes that

**Table 2.1.** Acronyms used in English teaching

| Acronym | Meaning |
|---|---|
| TEFL | teaching English as a foreign language |
| TESOL | Teachers of English to Speakers of Other Languages |
| TESL | teaching English as a second language |
| ELT | English language teaching/training |
| ESP | English for specific purposes |
| ESL | English as a second language |
| ESOL | English to speakers of other languages |
| EFL | English as a foreign language |
| EAP | English for academic purposes |
| IEP | intensive English programme |

**Box 2.1.** The tourist-traveller debate

The traveller vs tourist debate is one that has been ongoing for years. Travellers and tourists have inherently been defined and separated from one another and placed at two ends of an invisible travel spectrum.

According to a quote on the website Dangerous Business (2017):

> *tourists* are the ones who wear fanny packs and Aloha shirts. They're the ones piling out of big buses to strike stupid poses in front of all the tourist sites. They are the ones eating at McDonald's because they're afraid of 'weird' food. They are the close-minded ones worthy of ridicule.
>
> *Travellers* are the ones who tote backpacks and only 3 pairs of underwear around the world. They use local transportation and thrive on living in hostel dorm rooms. They eat at street stalls and interact with locals as much as possible. They are the adventurous ones worthy of envy.

In effect, there is a general presumption that tourists are 'bad' and travellers are 'good'.

in order to qualify as a tourist, one must travel and remain in a place outside one's usual residential environment for not more than one consecutive year for leisure, business or other purposes. Matheison and Wall (1982) however, do not impose a timeframe, simply stating that one must travel to a destination temporarily, while Leiper (1979) proposed his definition of tourism according to the three separate stances of the economy, technical aspects and tourism in a holistic sense. The first stance focuses on tourism as a business, the second on the tourist in order to provide a common basis by which to collect data, and the last stance attempts to include the entire essence of the subject.

While such attempts to define the concept may be useful from a generic perspective, the practical relevance of such definitions is difficult when applied to specific tourism types, such as TEFL tourism. Robinson and Novelli (2007), in their introduction to the niche tourism phenomena, theorise that tourists have developed as consumers, becoming increasingly sophisticated in their needs and preferences as a result of an emergent culture of tourism. Despite such acknowledgements of the progressive and adaptive nature of tourism, particularly evident through the limitless introduction of new and niche tourism forms of a micro- and macro- nature, there appear to have been no attempts to develop the commonly accepted definitions of tourism in parallel.

The debate over who is or is not a tourist may have been considered in a peripheral manner by the academic community in recent years, but this does not mean that it is not a centrepoint for discussion. There is considerable reference to the definition of tourist throughout the non-academic community, where many tourists claim not to be tourists, but instead wish to be classified as 'travellers'. While the description of a 'traveller' by many people (as demonstrated in Box 2.1) may align more closely with titles such as 'backpacker' or 'gap year tourist', it is important to note the differentiation in definitional perceptions between the academic and non-academic communities.

This gives rise to the concept of 'role ambiguity', which has been addressed by scholars investigating the volunteer tourism industry (in which unpaid TEFL is often included). Academics (e.g. Wearing, 2001; Stoddart and Rogerson, 2004; Lepp, 2008) have found that volunteers do not perceive themselves as tourists; indeed some actively promote an anti-tourist theme (Lepp, 2008). In contrast, some volunteers perceive themselves as a tourist at various points during their trip (e.g. when they have completed their volunteering duties) and as a volunteer at others (Sin, 2009). Mustonen (2005) similarly identifies the shifting roles of volunteer tourists who may go from being tourists seeking out pleasure, relaxation and stimulation during part of their trip to altruistically helping the community or environment at other times. Such ambiguities contribute to the complexities in defining not only the TEFL tourist, but tourism on a broader spectrum.

This book positions the TEFL teacher within the blurred boundaries of the tourist spectrum. There are two fundamental reasons for this association. The first is the movement of the TEFL teacher from their place of residence in their home country, to their temporary residence in their TEFL location. The discrepancy here, however, is that not all TEFL tourists will undertake trips less than a year in duration, as prescribed by the WTO and UN (1994). While this may be the most commonly cited definition by tourism scholars, it can be argued that this is unjustifiably the case, due to the definition's vague nature and simplicity, failing as it does to take into account the diverse nature of tourism in modern society. In addition, this definition does not consider the nature of temporary residency permits and frequent relocation of TEFL tourists, meaning that they may no longer sustain a 'usual residential environment', instead living a semi-nomadic life facilitated through undertaking employment in the English teaching sector.

The second justification for the association between TEFL and tourism is that many teachers, while residing in their TEFL location, undertake day trips or short breaks both within and outside their host destination. This can take the form of domestic tourism or international tourism. So, while it may not always be entirely clear whether the TEFL teacher qualifies as a tourist based on their overall teaching experience, it is likely that they do partake in the tourism industry during their leisure time.

Rooted in the ambiguous concepts of tourism and TEFL, the TEFL tourism phenomenon is subject to a significant level of complexity. Geographical and cultural contexts, surging globalisation, developing economies, the growing need for English teachers, the lack of regulation and the differences across education systems all contribute to the TEFL tourism industry being broad and diverse in nature. Taking this into account and in acknowledgement of these limitations, A TEFL tourist can be operationally defined as:

> A person who travels outside of their usual environment to teach English as a foreign language, whose role shifts between tourist, educator and educatee at various points in their trip.

The proposed definition is not intended to be representative of TEFL tourists across all localities and situations, but rather to act as a basis from which contextualisation can occur depending on the unique circumstances of different geographical locations. The notion of interchangeable roles is incorporated into the definition. This was derived from typological research (Stainton, 2017a) into the TEFL tourist, where it was identified that TEFL tourists frequently shifted between the roles of tourist and teacher according to the day of the week, geographic location and the people they were with.

**Box 2.2.** Domestic TEFL tourism

> Living in Thailand was a constant adventure. I worked hard during the week, but weekends and holidays were full of action. During my seven months I went jungle trekking and elephant riding on Koh Chang, chilled out on the beach on Koh Samet, visited numerous wats in Chiang Mai, watched Thai boxing in Bangkok, and experienced a full moon party on the beach in Koh Phangan. I thought about teaching in South Korea before I chose Thailand and I'm glad I didn't – it's cold and I wouldn't have had nowhere near as much fun as theres [*sic*] not as much to do!
>
> (Stainton, 2017d)

## A New Global Agenda

The limited academic reference to TEFL and its links with tourism to date is surprising given the extent and importance of the English language teaching industry worldwide. While there is ample evidence of the amalgamation of the two concepts of TEFL and tourism, as demonstrated throughout this book, the bulk of existing literature centres around pedagogical practice or development agendas. It is this gap in the literature which inspired the production of this book, with the ambition of raising awareness in both academic and non-academic communities of the benefits and pitfalls that the tourism element may present.

Global demand for English language abilities is continuing to grow throughout the world as the English language has become an integral part of world society. Governments are now increasingly recognising the importance of English to their economies and societies, and individuals often see English as a tool that can help them to fulfil their personal aspirations. The English language has become a centrepiece of educational reform in many countries, often in response to developments in economics, demographics and technology. Improving a nation's English literacy is not only important in terms of business, education and diplomacy, but also to enable the populace to enjoy global culture and entertainment. Therefore, there has emerged an almost insatiable demand for English teachers across the world. As of 2014, the number of English language learners worldwide stood at 1.5 billion and it is estimated that this figure will increase to over 2 billion by 2020 (British Council, 2014).

The lack of qualified English instructors presents one of the largest challenges to educators and citizens across the globe, and it is no surprise that many destinations are now looking towards native English speakers to fill employment gaps. It is estimated that 250,000 native English speakers work as English teachers in more than 40,000 schools and language institutes around the world (Teaching English as a Second Language (TESOL) International Association, 2014), although this figure may be significantly higher as a result of employment unaccounted for by the state, such as private tuition or those working without the correct visas or documentation. Training and recruiting such a large number of TEFL teachers across the globe is, of course, a challenging endeavour, putting additional strain on nations which may not always have the appropriate resources. In addition, educators have to plan for the inevitably high staff turnover. The TESOL International Association (2014) states that approximately 50% of TEFL teachers remain in employment in excess of one year, with 15–20% relocating to an alternative school or country after this time, 30–35% returning home and 10% continuing employment for a third year. This results in the need for global recruitment of over 100,000 TEFL teachers annually.

References to enhancing a nation's level of English speaking tend to be integrated into discussions regarding development agendas. This rhetoric is frequently encompassed within agendas associated with the UN's sustainable development goals (SDGs), where enhanced English is viewed as a means of improving education, driving economic growth and decreasing poverty. Appleby *et al.* (2002) differentiate between language **in**, **as**, **for** and **of** development. Language *in* development refers to the role that language plays in socioeconomic development and raises questions regarding the place of English and other languages and the way in which they may or may not contribute to local development. Language *as* development refers to the provision of language teaching and language projects where language provision is an end in itself. Language *for* development is language used as an important tool for the development of different aspects including business, science, law and media. Language *of* development differs from the preceding three as it refers to actual language used. It focuses on the lexico-grammatical and discourse properties of the English language while the other three refer more to the roles and functions of language.

While there are strong arguments for the link between language and development, there is equally strong evidence suggesting the link between tourism and development in current literature. However, this link is not yet complete. A good starting point is to place it within Appleby *et al.*'s (2002) spheres of language. On the basis that TEFL practices provide a platform for language teaching, it can be argued that TEFL, and by extension tourism, is in fact an example of language **as** development, therefore demonstrating an inextricable link between language, tourism and development (Fig. 2.1).

The relationship between the three concepts of language, development and tourism may not be apparent in existing academic literature, but this is in stark contrast to promotion material found on the internet when searching for TEFL opportunities. There is now a clear association between TEFL and the traditional 'package' style tours as evidenced by the purchase in 2007 of the volunteering company i-to-i by the profit-maximising organisation First Choice Holidays for approximately £20 million (Benson and Wearing, 2012). First Choice Holidays now promote paid TEFL opportunities at a cost. There are a wide range of organisations

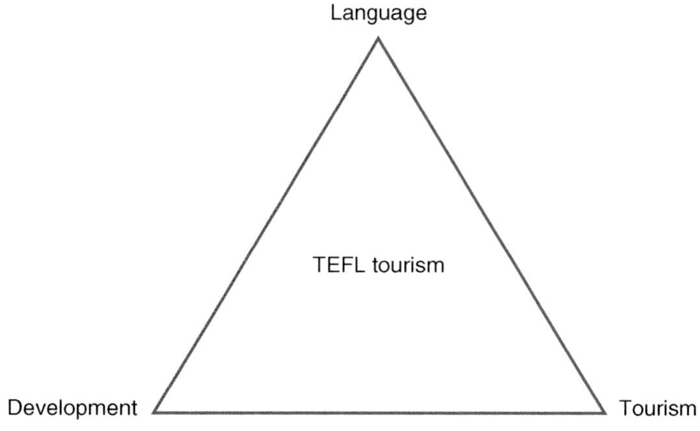


**Fig. 2.1.** The tourism, language, development relationship

> **Box 2.3.** Ghana teaching project
>
> Below is an excerpt from the website Frontiers (2017) which promotes opportunities to teach English in Ghana, demonstrating both the TEFL and tourism elements of the experience.
>
> > Experience a true Ghanaian welcome from the children of Ghana, surrounded by the natural beauty of this wonderful country, and truly integrate yourself into a friendly and hospitable culture. Teach underprivileged children and give them the opportunity to make something of their lives. In your spare time you could swim in waterfalls, spot elephants, baboons, crocodiles, and antelope on the African plains, or dance and socialise to the Ghanaian beats.

promoting opportunities which conjoin the two aspects of tourism and TEFL; however they have not yet been coined 'TEFL tourism'.

There are complex issues and challenges associated with this global movement towards English speaking, which are important to note. Such complexities are highlighted by the British Council which produced a publication (Coleman, 2011) to raise awareness of some of these issues. The text, entitled *Dreams and Realities*, draws together a range of voices from around the world, addressing issues including the perceptions of individuals and communities of the role and benefits of English; language policy and the introduction of effective change in education systems; the role of 'local Englishes' and their use alongside global standard English; the equality of access to English; and the contribution of English to national and international development. While analytical viewpoints such as this demonstrate good practice in the global education industry, there is no current analysis of issues such as those identified by the British Council when the practice of TEFL is combined with tourism, an industry renowned for its negative impacts when not carefully managed.

Perhaps the most important distinction to make is the difference between macro and micro approaches taken. Discussion of development in terms of TEFL tourism in this chapter so far has largely taken a top-down stance. Influenced by Foucault's notion of governmentality (Foucault, 1991), this macro approach looks at the actions and strategies of various agents involved in planning and development, whether these be politicians, advisers or educators. This approach has much in common therefore with the category of the language **of** development described above, as opposed to language **as** development as associated with TEFL tourism.

This suggests that the approach best taken is not from a macro perspective, but rather to assess TEFL tourism, its benefits and limitations, through a micro lens. The need for micro analysis and regulation has long been recognised in educational development as there has always been considerable interest in learners and teachers and their roles in the curriculum. However, this approach is not always adopted by all nations in all situations. There is a level of complexity added when taking into consideration the links between the macro and the micro. While there may be direct links in some instances, for example between government policies and formal educational implementation, in others the links may be less explicit or may not exist at all. Take commercial TEFL organisations for example. While these agencies may facilitate the practice of TEFL in a destination, is this likely to be aligned far more closely with their own capitalist profit-making agendas than with the nation's development goals?

## TEFL Opportunities Worldwide

English has been referred to as a 'barometer of western influence' to which almost all nations (with the exception of the likes of Myanmar and North Korea) have succumbed, to some extent. Globalisation brings an inevitable desire to learn English in order to be able to communicate with much of the rest of the world, in turn enhancing development and commercial prospects. This desire, therefore, requires large-scale recruitment of TEFL teachers. Employing such a large number of teachers, however, is not always an easy task and less favourable destinations are often subject to particular shortages (Box 2.4). With TEFL teachers in such high demand, it is therefore no surprise that English-speaking tourists are often recruited to fill the gap which would otherwise be left empty.

Opportunities for TEFL tourism are limitless, spanning all four corners of the globe. Griffith (2014), in her guide to teaching English abroad, attempts to demonstrate the scale of the TEFL industry, providing a summary of opportunities throughout the world. However, it can be argued that to provide an exhaustive list of TEFL tourist-receiving destinations is an impossible task due to the complexities of employment types and the available evidence. Statistical data are indispensable tools for sound evidence-based decision-making, planning, implementation and monitoring of any policies and programmes, including those related to TEFL. However, as emphasised by the United Nations World Tourism Organisation (UNWTO) (2014), data on tourism-related employment is fragmented, lacks quality and international comparability. This is the case not only at the international level, but also at the national level where different methods and sources of data collection often produce different figures and results.

**Box 2.4.** Teacher shortages in Malawi

Malawi has one of the world's worst teacher shortages which is equivalent to 2% of unfilled teacher posts globally. The result is that children are often squeezed into overcrowded classrooms, with those in early grades particularly disadvantaged. As of 2015 there were around 130 children per class in grade 1, on average, compared with 64 in the last grade. The student/teacher ratio in primary school increased from 63:1 in 1999 to 76:1 in 2011. The teaching force is growing at just 1% per year, which is far from sufficient to reduce the pupil/teacher ratio from 76:1 to 40:1.

(UNESCO, 2016)

Although enhanced employment prospects as a result of tourism has long been recognised, employment in the tourism industry has, up to now, been one of the least studied aspects. The tourism labour market has a dynamic nature – i.e. high labour turnover between organisations, wide range of remuneration levels and schemes, seasonality, etc. (UNWTO, 2014) and few countries have developed reliable methods of analysing employment in the tourism industry, largely owing to its diverse and dynamic nature. Aspects such as seasonality, part-time and/or excessive hours of work, low-paid (or unpaid) family labour, and informal or sometimes illegal labour makes measurement notably more difficult.

While measurement of the extent to which formal TEFL employment exists across the globe may be a challenging task, there is no reliable way to account for informal employment that may take place. The number of TEFL tourists employed will be underestimated when teaching takes place informally or when it is not legally registered. In addition, TEFL tourists may not always be formally acknowledged due to aspects such as the temporary nature of employment and residency, illegitimate visas and legal paperwork, avoidances of registered taxation and casual employment.

The definition of 'informal employment' is expressed differently by various organisations and researchers. The Organisation for Economic Cooperation and Development (OECD) (2015)

**Box 2.5.** Visas and working permit regulations in Thailand

It is a legal requirement for all foreign teachers to obtain the correct Thai visa. To undertake paid employment the TEFL teacher requires a non-immigrant B visa. To be granted this, the prospective teacher requires certified copies of their degree and TEFL certificates, a criminal record clearance, an official recommendation letter from the education institution in Thailand at which they will be based and information in relation to the contract or term of employment. Similarly, those undertaking voluntary work must obtain a non-immigrant O visa by providing official recommendation from the organisation in Thailand at which they will be based and information on the term of voluntary work. Both visas are single entry valid for 90 days.

Despite these guidelines, presented on the website of the Royal Thai Embassy (2016), Methanon-pphakhun and Deocampo (2016) in their research found that visa requirements were a major source of frustration for TEFL teachers as a result of long visa runs and confusing rules. Stainton (2017a, b, c, d) in her research found that 45% of TEFL tourists were at some point during their employment working without the correct legal documentation.

**Box 2.6.** Illegal voluntary employment in Thailand

Research investigating TEFL teachers in Thailand indicated that a number of teachers were unaware that they were working without the legal documentation. Voluntary positions require a non-immigrant O visa. However, frequent remarks indicated a misunderstanding in this regard. One respondent in Stainton's (2017d) research explained:

> I was in Thailand as a volunteer teaching tsunami survivors English. Because I was a volunteer I didn't need any paperwork.

This respondent appeared to have been formally employed, but was not legally registered.

and the International Labour Organisation (ILO) (2012) state that the employment which is not included in the evaluation of the overall employment quantities of the population of a country or destination in an official regard qualifies as informal employment. Typically, these activities are carried out without the conclusion of an agreement, without taxation and in violation of legal regulations. The European Union Committee, however, does not prescribe that informal employment is inherently illegitimate, rather activities are regarded as informal through failure to report to the authorised organisations (Vermeylen, 2008). In the tourism industry specifically, informal employment has been known to include the non-declaration of employees to the public authorities and institutions; willingness of employees to work without insurance; reluctance of employers to declare low wages and working hours (which may breach legal limits); appointing interns with tasks that exceed the educational purpose; and extensive labour (Çelik *et al.*, 2016).

The informal economy comprises half to three-quarters of all non-agricultural employment in developing countries (ILO, 2012). Given the inherent links between TEFL, tourism and development, it is evident that TEFL tourism activities are populous throughout the developing world and therefore highly likely that the informal economy workforce comprises tourists teaching English as a foreign language, although the extent to which this is so is, of course, unknown. TEFL tourism is no exception to the struggle of obtaining reliable employment data and therefore to quantify the industry would be an impossible task. TEFL tourism industries can be characterised by diversity both on the basis of intra-national and international criteria causing the nature of employment in the industry to vary significantly. The range of sub-sectors, the size of organisations, their ownership and the markets they serve demonstrate that the nature of English teaching varies substantially. As a result, it has become increasingly necessary to take a broad view of the labour market and consider its close links to its allied labour markets of education and tourism.

While the definition of formal and informal employment may not be clear-cut, it can be suggested that tourists who are employed to teach English overseas will fall into one of three categories. The first is formal employment, where the teacher is employed in accordance with local administrative requirements and regulations. This includes working in an official capacity in places such as government or private schools, tuition centres or language institutions. These teachers are likely to be in possession of an employment contract, pay any relevant taxes and be represented in national employment figures. Paradoxically, TEFL tourists who are employed on an informal basis are not likely to have recorded regular teaching commitments and they may undertake work such as home or one-to-one tutoring. These tourists may not hold the correct paperwork to be legally working in the country and are unlikely to be represented in national statistics. While there may be an expectation that the former undertake legally registered employment and the latter do not, it is important to recognise that this is not absolute. A teacher may work in an official regard at a school, but not hold the correct visa and therefore be formally employed though not legally registered. Similarly, a teacher may be formally and legally employed and choose to work as a private tutor, declaring any monies earned, paying the correct taxes and keeping up to date with relevant legal administrative processes.

The nature of TEFL varies significantly across geographical locations, educational institutions and employers or placement organisers. The lack of regulation of the TEFL industry heavily contributes to the diversity and range of opportunities available to the TEFL tourist. Positions may be based either in the teacher's home country or abroad, can involve varying levels of teaching and can include both paid or voluntary work. Placements can be self-sourced or organised through the use of an 'agency', such as a charity or for-profit tour operator.

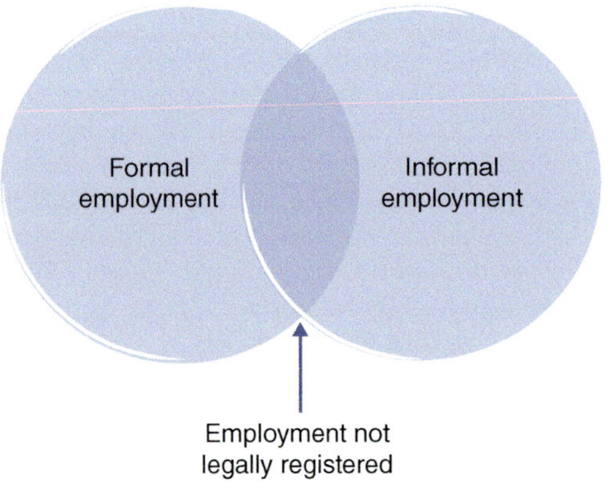

**Fig. 2.2.** Employment types in TEFL tourism

## Voluntary vs paid positions

Perhaps the biggest distinction to make within TEFL tourism employment is whether the employee is remunerated or if they are working on a voluntary basis. This adds another layer of complexity when examining the TEFL tourism industry, considering the inherent differences between aspects such as teacher motivation, skillsets and length of placements. Should employment be unpaid, the TEFL teacher would likely qualify as a volunteer tourist, an industry that has been subject to a vast array of research and critique in recent years. Should employment be paid, the TEFL teacher may be classed as an expatriate, paid intern or temporary worker. These differing associations further highlight the diversity of the TEFL tourism industry.

Deeming whether employment is voluntary or otherwise, however, is not as clear-cut as it seems. In fact, there is no universal definition of volunteering (Volunteering England, 2008), largely owing to the term's subjectivity. This subjectivity is particularly noted when comparing perceptions of what constitutes a voluntary activity across different geographical regions pertinent to the study of TEFL teaching, owing to its global nature. Further to this, the blurring of paid and voluntary work has become commonplace (Lyons, 2003; Lyons and Wearing, 2012) and it is now typical for organisations to provide incentives such as reimbursements, cash and qualifications as part of the volunteering experience. This makes it unclear whether participants are in fact paid employees or unpaid volunteers.

Both voluntary and paid positions tend to derive from one of three broad avenues. The first is self-sourced employment, where the TEFL teacher locates the position themselves, whether this be through advertised positions or speculative application. The second and third involve the use of a third party to organise the TEFL position. Historically, development organisations such as Voluntary Service Overseas (VSO) and the Peace Corps along with charities such as Oxfam, populated the market for volunteer teaching overseas. Nowadays, however, postmodern, capitalist society has given rise to an increasing number of commercial businesses which align themselves more closely with a traditional tour operator than with a development organisation whose ethos is integrally linked to concepts of philanthropy, altruism and goodwill.

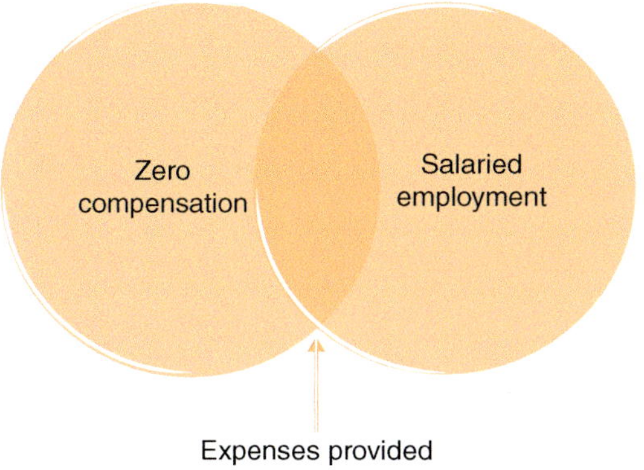


**Fig. 2.3.** Remuneration types in TEFL tourism

### Self-sourced positions

There are various ways of securing employment independently, ranging from speculative applications and word of mouth, to recruitment agency assistance and internet applications. The extent to which one may be successful in locating TEFL positions depends perpetually upon a number of variables. Some destinations may have formal advertising and application processes, whereas others may rely largely on informal applications and word of mouth. Finances play an inevitable role here, dictating the extent to which an employer may be able to promote their vacancies and subsequently affecting the extent and quality of the pool of applicants. A government-based school in Zambia, for example, is unlikely to be able to afford to advertise their position in the *Guardian* newspaper, whereas a wealthy school in the United Arab Emirates (UAE) might (Box 2.7). Likewise, some educational establishments may source applicants through the use of a third party, such as a recruitment company, where the institution pays a fee for their services.

This does, of course, exclude much of the TEFL community throughout the developing world, where budgets and resources may be limited. In many areas, self-sourced employment is likely to take place through pre-established

**Box 2.7.** English language teacher vacancy in the UAE

English language teachers – UAE

Worldteachers are working closely in partnership with the UAE education reform project. They are currently looking for several experienced English teachers to teach within the state school system in the UAE.

**Minimum job requirements**

- Bachelor's degree in English preferred.
- CELTA/Cert TESOL/(or recognised Level 5 equivalent)/or English degree is an advantage.
- Minimum of 2 years teaching experience in the EFL/CELTA field.
- Excellent oral and written English essential (IELTS 7.5 Academic for non-native speakers).

**Salary**

Total salary is AED 21,000 per month which is the total figure paid and includes living allowances. Other benefits paid include flights and medical cover.

Applicants must be able to attend a face to face interview in UK or UAE during October 2017.

Worldteachers will help you throughout the recruitment process and be able to provide full information on the opportunity.

(*Guardian* online, 2018a)

contacts or word of mouth. While this may not require the outlaying of significant costs, the organic reach will inevitably be far less than if external marketing methods were to be employed. While it is impossible to generalise across the globe, the assumption can be made that the less wealthy the employer, the less capital is invested in marketing and recruiting, thus leading to a lower quality of applicants for the position.

One solution to this problem is for the educational establishment to feature on job forums or directories. While some of these may require a subscription fee to be paid, others may provide the perfect opportunity for free advertising and promotion. As the internet is the first port of call for many job seekers nowadays, websites, social media and search engine optimisation (SEO) are fundamental to achieving an online presence. Websites can be created and managed for little cost and forums and social media platforms provide the opportunity for employers to promote their positions free of charge. While, of course, there is a prerequisite that there must be adequate internet access – an issue for concern in some parts of the world – this provides ample opportunity for organisations that may previously have been hindered by budget to reach audiences that were formerly unobtainable.

There are a number of websites which host TEFL industry information and data, including job vacancies. Many of these are cross-national, such as the ESL Café or Total ESL and include information spanning different geographical regions. Others are destination-specific such as Ajarn in Thailand.

## Development agency positions

Traditionally, the main avenue to source employment as a TEFL teacher in a developing country was via a development agency, also referred to as an aid agency. These agencies can exist, both within government (e.g. AusAID, USAID, DFID, EuropeAid), between governments as multilateral donors (e.g. UNDP) and as private voluntary organisations or non-governmental organisations (NGOs) (e.g. ActionAid, Ducere Foundation, Oxfam, World Vision). Their purpose is to promote economic growth and development in the areas they serve. Aid can be subdivided into two categories: humanitarian aid (emergency relief efforts, e.g. in response to natural disasters) and development aid (or foreign aid), aimed at helping countries to achieve long-term sustainable economic growth and poverty reduction. It is the latter which concerns TEFL tourism.

Opportunities with such organisations nowadays, however, are not as abundant as they once were. Development organisations and charities often carefully vet prospective TEFL teachers, who are almost exclusively recruited on a voluntary basis. Commercial volunteer agencies offer short-term placements to people regardless of qualification and/or experience, whereas development or aid agencies tend to seek qualified and experienced people to work overseas on paid fixed-term contracts. Locating a position with a development agency is a challenge in itself, with the market now saturated with capitalist enterprises more closely aligned with the tourism industry than international development. Many of the organisations above no longer send volunteers overseas, instead concentrating their efforts on financial assistance and offering voluntary fundraising opportunities as a means of helping developing nations as opposed to hands-on assistance in-country.

VSO is the key player in the development field nowadays, with education-based programmes in 17 countries across the world. VSO has been operational for almost 60 years and has supported almost 40,000 volunteers to date. It can be argued that development agencies, such as VSO, have a better knowledge of, understanding

**Box 2.8.** English language trainer, Myanmar

Education in Myanmar is often conducted using the English language. However many of the teachers employed do not possess the level of English that they need to educate their students to a necessary proficiency. Alice Redfearn is a VSO volunteer. She works with the British Council to deliver English Language training to the lecturers of student teachers, combining strong English language skills with good, reputable teaching models. The programme is in place in a range of universities and colleges working throughout the country which aim to build up the capacity of next generation of teachers.

(VSO, 2017)

of and desire to assist local communities than their commercial counterparts. Instead of simply sending teachers to areas in need to fill employment gaps, VSO recruits volunteers to support the training of teachers. This approach aims to encourage a self-sustaining education industry in the future, where teachers in-country can train future generations, thus reducing the need for TEFL teachers to be imported from western destinations in the way they are now.

There are arguments to be had both for and against using development agencies to enhance a country's English language capabilities. Within the foreign aid sector, the overarching goal is to improve livelihoods while reducing inequality and injustice in developing countries. There is also the opportunity to create strong bonds with other countries and this can be beneficial for not only the country receiving the aid, but also the giver of financial aid, allowing for diplomatic relations to be formed and nurtured. On the other hand, there are opponents. Some countries are run by corrupt leaders and officials who choose to allocate resources according to their own personal gain rather than prioritising the benefits to the community. Furthermore, some countries that give aid may use the donation as a tool to control the recipient country in exchange for diplomatic favours such as setting up military bases.

### Tour operator positions

In the UK, all TEFL agencies will be registered companies. Many of them are businesses which make a profit; others are registered as charities. Charities are not allowed by the regulatory body to make unnecessary profits. Other TEFL organisations may be working to a non-profit model (but not registered as a charity) and some may be social enterprises. Some TEFL companies are large and profit-focused, with overseas offices and representatives just like a travel company; others have no in-country staff or local support mechanisms. Without doubt, organisations operating within the TEFL industry are diverse, with differing types of ethos, motives and operational ideals, and it is clear that the commercial sector is becoming prominent.

Promotion of the TEFL industry is no longer littered with rhetoric of philanthropy and goodwill, but instead cites fun, skill development and subjective notions of 'finding oneself'. Research is easily conducted using the internet and a search for the term 'TEFL teaching overseas' in Google yields over 428,000 results. Most of the top listings are from businesses with considerable advertising budgets while many of the small charities or social enterprise agencies appear further down the list or not at all. These agencies largely promote short-term TEFL placements at a cost to the teacher and often include additional incentives such as qualifications, transport arrangements, accommodation and excursions.

According to the United Nations, a tour operator is a business which combines two or more travel services (e.g. transport, accommodation, meals, entertainment, sightseeing) and subsequently sells them through travel agencies or directly to final consumers as a single product for a global price. The components of said package tour might be pre-established or can result from an 'à la carte' procedure, in which the visitor decides the combination of services he/she wishes to acquire. Traditionally this model has been associated with the common package holiday, consisting of accommodation and transport arrangements. However, in a postmodern tourism industry this has become invariably obscured in order to achieve the best-fit model for the product or service sold.

In today's context of TEFL tourism, there is a clear alignment between commercial TEFL

**Box 2.9.** TEFL Heaven TEFL tourist programme in Costa Rica

TEFL Heaven offers a programme in Costa Rica entitled 'Costa Rica: Manuel Antonio Beach + Job'. The highlights advertised are:

- 4 weeks face-to face TEFL training
- Paradise location
- Qualifies you to teach English anywhere in the world
- Guaranteed paid job
- Live and work in one of the most popular countries
- Mix with like-minded people
- Accommodation assistance
- Airport pick-up
- Free salsa lessons

Fee: £1315
Project duration: 6 months +

(TEFL Heaven, 2017)

agencies and tour operators. The company 'i-to-i' is an excellent example of the postmodern tourist packaged product, offering TEFL placements combined with various elements including accommodation and employment assistance (Box 2.10). It is owned by the profit-maximising organisation First Choice Holidays which is typically associated with package holidays. This provides further indication of the links between the package-based tour operator and TEFL tourism provision.

Tour packages and many tour operators tend to be associated with the mass market and all of the sins which accompany that. If we are to view the TEFL product as a postmodern tour package as opposed to a predominantly altruistic endeavour, then we may be able to learn lessons from the tourism industry, allowing for the potential mitigation or reduction of any associated negative impacts. The ambiguous nature of TEFL organisations, coupled with the lack of regulation of the wider TEFL industry (Griffith, 2014; Kogar, 2014), gives rise to the question of whether those organisations with predominantly financial motivations will recruit the most suitable teachers to help enhance the country's social and economic development and train them in the best way. As pointed out by researchers investigating areas similar to TEFL tourism such as the volunteer tourism market (Tomazos and Butler, 2009; Keese, 2011), profit-focused organisations have the potential to exploit local communities and TEFL teachers as opposed to helping them.

Deciphering whether a TEFL organisation is development-focused or profit-orientated, however, is no easy feat. In the current market, there is no clear distinction between the types of agents offering TEFL placements and the differences in their services and values. This is akin to the volunteer tourism industry where it is argued that the ever-evolving marketplace contributes to an ambiguous industry. Like the volunteer tourism marketplace, variations in size, ethos and businesses cause concern with regard to the value of projects promoted by these agents. The current lack of regulation of the industry leaves the door open for opportunists and it is imperative that this is observed and monitored by stakeholders in order to ensure sustainable management of the TEFL tourism industry.

**Box 2.10.** i-to-i TEFL tourist programme in Peru

i-to-i offers a TEFL tourism programme in Peru entitled 'Teach English in Peru!' It promotes the highlights as follows.

- Discover a new culture and learn Spanish in a program endorsed by a professional and serious organization
- First-week accommodation is covered by the school
- Teaching job in Lima or in another main city
- Pre-departure assistance (visa application process)
- Welcome document: tips, advice, and practical tools
- Health insurance (health care coverage in Peru)
- Orientation in Lima
- Emergency support while in Peru

Fee: £1125
Project duration: 1 month

(i-to-i, 2017)

## Ambiguities in TEFL tourism

One of the reasons for the production of this book is to highlight the ambiguities in the industry which make it extremely difficult not only to regulate, but also to comprehend aspects such as scale and quality. While there may be data available regarding English language learning in a particular locality, this certainly does not apply across the board and there is little standardisation. TEFL organisations pose one of the greatest problems in this regard, with differences in ethos, motivation and operational practices affecting the shape of the industry. Geographical and financial restrictions add further layers of complexity to the understanding of the TEFL industry, along with employment types and the positions offered to TEFL teachers.

Perhaps the biggest challenge facing the TEFL industry globally is the issue of subjectivity. Perceptions of TEFL and how best to manage it inevitably differ according to a person's prior knowledge and experiences. This notion of ethnocentrism is fundamentally underpinned by geographical and cultural differences and, it can be argued, makes it an impossible task to achieve standardisation. While a globalised approach in this way may not be the end goal, the mixing of educational backgrounds and viewpoints provides scope for further perplexity.

Most studies addressing the TEFL industry (or EFL, ESL, etc.) take a pedagogic perspective, addressing the likes of teaching practices, methods, assessments and enrichment. There is a considerable gap in the current literature, however, in terms of TEFL as a social, corporative or fiscal endeavour. While this book intends to fill this gap to an extent, to provide a comprehensive overview would be an impossible task. Most notably, the lack of formal regulation within the industry contributes to this hazy comprehension of TEFL tourism. Other areas of particular concern include the standards of training for prospective teachers, the variety of TEFL qualifications available to the consumer and their worth, and the skills and suitability of staff employed.

## Regulation of the Sector

Currently, the TEFL industry is notably unregulated. There is little clarity over the job description and person specification of a TEFL teacher, nor are there any standardised processes with regard to contractual agreements, pay scales, rights or working conditions. It is aspects such as these that are the precursor to the grey dynamics currently presented within the TEFL industry on a global scale. While to a certain extent this can be deemed inevitable due to the complexities noted in the previous section, it can be extremely problematic for stakeholders involved with TEFL practices and the TEFL community. A lack of regulation can significantly and inversely contribute to the sustainability of the TEFL tourism industry and thus it is imperative to bring said issues to the forefront of the English language teaching discussion.

Essentially, TEFL is an umbrella term for teaching English to students of any age whose first language is not English. Beyond this, however, specific duties can vary considerably. According to Prospects, a UK-based charity which provides guidance for graduates and jobseekers, a TEFL teacher might use a variety of audio-visual aids (Box 2.11). However, this may not be possible when teaching in institutions which do not have

**Box 2.11.** TEFL teacher job description

According to Prospects, the following is expected of a TEFL teacher.

> TEFL teachers use a range of course books and materials, plus a variety of audio-visual aids. A strong emphasis is placed on dialogue and role-playing, but more formal exercises, language games and literature are also used.
>
> The content of lessons varies depending on the reason why the students are learning English, e.g. whether it's for business use for adults, school work for children and so on. The aim of each lesson is to encourage the students to communicate with each other using the structures and vocabulary they've learnt and to improve the four basic language skills of listening, speaking, reading and writing.
>
> Typical tasks that may be carried out include:
>
> - classroom management;
> - planning, preparing and delivering lessons to a range of classes and age groups;
> - preparing and setting tests, examination papers and exercises;
> - marking and providing appropriate feedback on oral and written work;
> - devising, writing and producing new materials, including audio and visual resources;
> - organising and getting involved in social and cultural activities such as sports competitions, school parties, dinners and excursions;
> - attending and contributing to training sessions;
> - participating in marketing events for the language school;
> - preparing information for inspection visits and other quality assurance exercises;
> - freelance teaching on a one-to-one basis;
> - basic administration, such as keeping student registers and attendance records.

(Prospects, 2017)

access to such facilities. In some countries, the TEFL teacher may not be required to undertake classroom management; instead there may be a local teacher to do this. Some countries may present a strong focus on cultural activities, such as taking part in school productions, whereas others may have rigid schemes of work to adhere to. Some schools may provide curriculum, resources and schedules and others may expect the TEFL teacher to create resources themselves.

While the duties required of the TEFL teacher may be somewhat ambiguous and may have subsequent impacts on aspects such as retention, job satisfaction and productivity, this is not really the main concern when considering the need for regulation. The TEFL tourism industry appears to have resisted critical scrutiny at large, owing to its laudable character and limited research, thus resulting in what could perhaps be viewed as a somewhat utopian sector of the tourism industry. The negative impacts resulting from a lack of regulation in TEFL tourism are far ranging and have not been adequately documented or examined within current industry practice or academic literature. Lessons can be learned, however, from the wider tourism community, where a lack of regulation has been a contributing factor to the rise of a number of negative impacts on both the host community and the tourist.

Such impacts have been explored within the volunteer tourism industry, in which TEFL is frequently incorporated. Coren and Gray (2012),

**Box 2.12.** Reform and regulation in EFL

Excerpt from the *Korean Times*, Wedneday 3 January 2007: EFL sector needs reform and regulation, not blame.

> I have taught in Korea since 2002, I have noticed a real hardening of the Korean public and government's attitude towards our profession and its place in this society… It seems like hardly a week goes by without some fresh scandal embroiling the foreign EFL community. Drugs, alcohol, sex, forged degrees, visa violations, unlawful public performances, internet stalking/addiction, an assortment of workplace problems, the occasional assault and a general inability to learn Korean thrown in for good measure all contributing enormously to making teachers an easy target for ridicule and scorn…
>
> Once considered a highly desirable international EFL destination, the government's lack of regulation has allowed the position of native speaker English teacher in Korea to become increasingly marginalized. Our rights and working conditions have been eroded to the point where no qualified, competent and halfway respectable foreigner in their right mind would give up life back home to ply their trade in this country. How long will it be until English teaching is added to the list of undesirable positions which make up the so called 3D (dirty, dangerous and difficult) job list filled by immigrant workers from the poorer regions of Asia?
>
> Instead of blaming native speaker English teachers working in this country for an employment visa system with more holes in it than even the thinnest slice of Swiss cheese, why doesn't the government clean up its own backyard by introducing a set of guidelines which are administered consistently and evenly across the nation? Following this, they could put a leash on recruiters and language institutes by implementing reform and regulation which would see the end of jobs where some teachers complain of suffering abuse and harassment akin to the horror stories experienced by 3D workers in the manufacturing sector.
>
> The reform and regulation of Korea's EFL industry would raise the bar for all parties involved and result in better jobs and working conditions for native English speaker teachers. This would attract higher calibre applicants with language teaching-specific tertiary qualifications who are currently choosing far more lucrative destinations in terms of money, working conditions and quality of life abroad. If the government demonstrates political will and suffers the initial pain of reform, they will have an industry which appeals to professionals and helps increase the country's low English proficiency skills to internationally competitive levels. Choose to maintain the status quo, and they will attract more of the wrong type of people with the day not too far away when native English speakers will no longer be willing to teach on the peninsula.

(ESL Café, 2016)

Gray and Campbell (2009), Coghlan (2006) and Palacios (2010) all identified in their research a feeling of unhappiness among volunteer tourists as their experience did not meet their expectations. While there are a range of laws in the area of advertising and consumer protection, the global nature of the industry means that this is difficult to police across borders. It can also be argued that such laws hold a certain amount of subjectivity where aspects such as deceptive messages, important information or aggressive sales techniques can be perceived differently by stakeholders and individuals.

Tomazos and Cooper (2012) identified a further negative impact in their research where they found reasons why the need for volunteers is sometimes simulated. This poses one of the most significant and controversial problems in volunteer tourism, where the profit-focused sending organisation arranges volunteer trips at the expense of the local community in order to make money as opposed to serving the needs of the local people. Within many areas of the volunteer tourism industry this can be easily demonstrated. Frederikke Lindholm from The Shelter Collection in Vietnam (Purvis & Kennedy, 2016), for example, documented that school trips have been undertaken where students build houses during the day and local builders work during the night to straighten the walls due to a lack of skills possessed by the volunteers. Another example, which has been well acknowledged in the media, is of orphanages in Nepal, where a number of poor parents are routinely duped into sending children to homes where owners use them to extract money from foreign visitors. Assessing whether TEFL teachers are needed or whether they are indeed exploited in ways that are evidenced above, however, is less clear-cut.

While it may be appropriate to justify the need for a TEFL tourist, particularly in an area where there is evidence of the need for English language development, the TEFL teacher may not provide the desired outcomes as a result of a lack of structure, guidance and regulation. Surprisingly, there is little evidence of this in the media outside of orphanages, but with short-term placements and without set teaching criteria, it would be very difficult to avoid negative outcomes such as teaching repetition and maintaining a consistent learning programme for the students. A neglect of locals' desires is a well-documented impact within studies addressing the volunteer tourism industry (e.g. Matthews, 2008; Guttentag, 2009; Benson and Wearing, 2012) and it can be argued that repetitive teaching and the employment of teachers who may not be adequately skilled demonstrates a sense

**Box 2.13.** Repetitive teaching in Ghana

Sandy organised a one-month TEFL teaching placement in French Guiana as part of her gap year. She wanted to 'do something different' and to 'help those less fortunate than herself'. On her first session she taught the group of students (aged 9–10) to say 'hello', followed by 'how are you?'. She then recited the numbers 1 to 10 with her students. At the end of the session she felt proud of the work she had done and genuinely thought that she had helped the children to learn English, an opportunity that she believed they would not have should she not be volunteering in the school.

Little did Sandy know that James, the previous volunteer, had covered exactly the same content. So did Kathy, the lady who worked there the month before James, and also Jemma who was there the month before that. In fact, the students should have been far further in their English language development than they were, given the number of teaching hours that they had received, but they were not because, rather than advancing through a curriculum and building on their knowledge, the teachers covered the same content that the students had already grasped.

Sandy was also wrong about the demand for her skills. Little did she also know that there were several English teachers who were native to French Guiana who had been made redundant in exchange for the free volunteer labour provided by the volunteer tourists. As a result of the TEFL tourism industry, these teachers were now unemployed and displaced. They continue to feel considerable resentment towards the TEFL tourists who, despite their philanthropic intentions, have caused problems among the local community. Furthermore, student grades have not increased since TEFL tourists have begun working at the school, but have in fact declined as a result of a lack of direction within the curriculum and assessments.

**Box 2.14.** TEFL in China: job advertisements

English First (EF) recruits all year round with new EFL jobs in China every month. With over 200 schools in 60 cities, EF is by far the largest recruiter of ESL teachers in China. It doesn't matter if a person is new to teaching, or if they have lots of experience. Here is a summary of the jobs advertised at the time of research.

Shanghai, English Teacher, EF English First Kids and Teens schools, eight positions

Beijing, English Teacher, EF English First Kids and Teens schools, six positions

Beijing, English teacher, EF English First Adult Centres, four positions

Guangzhou, English Teacher, EF English First Kids and Teens Schools, seven positions

Guangzhou, English Teacher, EF English First Adult Centres, two positions

Shenzhen, EF English First Kids and Teens schools, six positions

Chongqing, English Teacher, EF English First Kids and Teens Schools, three positions

Many other cities across China.

(English First, 2016)

of neglect. This is an area which requires urgent scrutiny within the TEFL industry.

Another area warranting regulation is that of pay. TEFL salaries vary widely from country to country. Some companies offer incentives like accommodation and the cost of return flights, while others provide living expenses and health insurance costs. This varies greatly across the sector. Some countries, particularly western developed nations, have their own regulations in terms of minimum pay. This is not, however, replicated in all parts of the world, particularly in destinations where employment rights are not a priority in the political agenda. Of course, there are vast differences in typical salaries offered between destinations, and while one might expect this to be due to the differing economic states, costs of living and average salaries in the various destinations, this does not appear to be the case when analysing some of the available data online. In fact, there are vast differences between the amounts paid in different geographical locations compared with the average salary in the same destination.

As a general rule of thumb, it seems that developing countries tend to pay TEFL teachers more than the average wage in the country, whereas developed nations tend to pay less. This is perhaps a reflection on the way that a TEFL teacher is perceived across different nations and the importance placed upon the position both in terms of teacher shortages and the country's development agenda and English language priorities. A further reason for this could be the differentiation between TEFL teachers and English teachers. In developed countries, a TEFL teacher would not be responsible for teaching English classes as part of the national curriculum, whereas in many developing nations this may constitute the primary focus for the teacher employed, where there is no difference between a 'TEFL teacher' and an 'English teacher'.

Although the employment rights of those in the western world will typically mean that employees are entitled to a certain level of pay or minimum wage, it appears that there could be issues at the other end of the spectrum in developing nations, where the employee rate of pay exceeds the national average. Payment inequalities can give rise to a number of negative consequences including displacement, gentrification and resentment. It is important to recognise that without the relevant legislation, it is possible for TEFL teachers not only to be underpaid, but for said employees to be paid an amount which may not appear reasonable to the indigenous population.

Countries have their own system of laws intended to protect the interests of workers and assure a minimum standard of living for the population at large. Regulation in this regard will, of course, differ considerably between nations and there tend to be more stringent employment rights for those employed within the western world than for those working in a developing

**Table 2.2.** Average TEFL salaries compared with average in-country salaries*

| TEFL tourism destination | Approximate rate of pay (USD) per month | Approximate average salary in-country | Difference between average TEFL salary compared to average salary in-country |
|---|---|---|---|
| United Arab Emirates | 4000 | 3500 | +500 |
| USA | 3500 | 3700 | –200 |
| Saudi Arabia | 3000 | 2700 | +300 |
| New Zealand | 3000 | 4000 | –1000 |
| Japan | 3000 | 2700 | +300 |
| United Kingdom | 2700 | 3000 | –300 |
| South Korea | 2500 | 2200 | +300 |
| Vietnam | 2000 | 150 | +1850 |
| Italy | 2000 | 2400 | –400 |
| China | 2000 | 850 | +1150 |
| France | 1800 | 3500 | –1700 |
| Spain | 1300 | 2600 | –1300 |
| Thailand | 1000 | 400 | +600 |
| Hungary | 1000 | 800 | +200 |
| Brazil | 1000 | 650 | +350 |
| Cambodia | 800 | 300 | +500 |
| Egypt | 800 | 550 | +250 |
| Turkey | 800 | 800 | 0 |
| Costa Rica | 700 | 550 | +150 |
| Poland | 700 | 800 | –100 |
| Mexico | 700 | 450 | +150 |
| Bolivia | 500 | 250 | +250 |
| Colombia | 500 | 400 | +100 |
| Peru | 400 | 650 | +250 |
| Venezuela | 300 | 200 | +100 |

*Data based on TEFL programmes currently offered and promoted and statistics available on the internet. This data is for illustrative purposes and is indicative only. It has not been methodologically verified.

country. Employment laws essentially govern the individual employment contract, providing a level of protection both for the employer and the employee. Across the western world, collective or industrial relations laws regulate the development and enforcement of collective agreements, the running of trade unions and any industrial action taken by workers and employers. Alongside this, social security laws govern social responses to any needs and conditions that are likely to have an impact on quality of life including old age, disabilities, sickness, death and unemployment.

The issue here lies within the geographical nature of the TEFL industry and the way in which a person's ethnocentric perceptions of the teaching experience and the education industry may be based upon their experiences within one area of the globe, but their TEFL placement is located in a completely different part of the world. While it would be naïve to make sweeping judgements of prospective TEFL teachers, it is fair to say that there will likely be a substantial portion of them who are not familiar with the rights that they will or will not have or expect in their TEFL destination.

In terms of the screening of participants, TEFL organisations go to different lengths in order to ensure 'quality'. One particularly important aspect of this is whether or not those employed to teach English as a foreign language have a criminal record or history (Table 2.3). International background checks are complex and notoriously difficult to carry out. Since there is no global database that allows schools to look at a teacher's information in one go, international background checks should consist of checking records in each country where the teacher has lived and taught. To complicate matters further, each country has its own set of laws, and some countries prohibit any agencies from obtaining records.

**Box 2.15.** 'Hidden' employment laws in China

According to an online article (Scam.com, 2018), most of China's 29,879 foreign teachers are not aware of China's labour laws that give foreign expatriate workers in China the same employee rights as Chinese. It appears that Chinese employers and recruiters want to keep this general ignorance in place since it is already difficult enough to find and recruit qualified teachers for China. The inconvenient truths of these powerful rights is simply not good for the recruitment business since it exposes such things as illegal contracts, and how expats have the right to walk away from such contracts without penalty.

Recent times have seen the Ministry of Labour and the Courts of China ruling in favour of foreign teachers who have filed formal complaints and/or lawsuits and have prevailed over three large chain operations including Disney English, English First and Webb. Teachers who find themselves being exploited are entitled to file formal complaints and recent cases include employee settlements of up to US $90,000. The following abuses are illegal (although remain to be commonly seen within the workforce).

- Unpaid overtime
- Being used as a marketing 'monkey'
- Having your personal image used in promotions
- Having your personal information disclosed to third parties
- Not getting a copy of your original chopped contract in English
- Not knowing the ownership identification of your employer
- Not knowing the SAIC business licence number of your employer
- Not being provided with a written job description
- Being told to lie to customers about your citizenship or credentials
- Being given illegal probationary periods exceeding 30 days
- Being forced to sign illegal contracts
- Not given 72 hours advance notice of work schedule changes
- Not provided with the proper Z work visa required by law
- Being forced to pay money for your release letter or Z visa
- Not being provided medical insurance as a full-time employee
- Not being paid within 48 working hours of a regular payday
- Illegal deductions and 'deposits' withheld from your pay
- Denied the right to see your employment tax records and receipts

(Scam.com, 2018)

While checking a person's background may not be an easy feat, there is plenty of evidence to demonstrate the importance of this. Most notably, children are frequently naïve and vulnerable and TEFL teachers are placed in a position of 'trust'. Broadly speaking, a position of trust can be described as a relationship within which one party is in a position of power or influence over the other by virtue of their work or the nature of their activity. Recent years, however, have seen an increase in people abusing their position of trust which has resulted in a wide number of public cases recorded, covering everything from accounts of theft to sexual abuse.

Recent headlines have placed heavy emphasis on the notion of sexual exploitation, with good reason. It is difficult to know exactly how many people worldwide are victims of commercial sexual exploitation, but there is ample evidence to suggest that the number is increasing. Accurate data concerning child sexual abuse is difficult to collect due to differing legal definitions of abuse and issues of under-reporting, with less than 50% of all sexual assaults on children being reported (Floric *et al.*, 2012). Most abusers are often known to the child and are 'trusted adults', and there have been a number of recorded instances where teachers have abused their position and engaged in sexual relationships with their students. While there are stringent laws governing such activities and rigorous background checks in attempt to avoid such occurrences across many parts of the world, there are also many destinations where such efforts are not employed.

To put this into context, research undertaken by Tomazos and Butler (2009) found that

**Table 2.3.** TEFL teachers and criminal records

| Country | Will not accept those with a criminal record | May accept those with a criminal record | Will accept those with a criminal record |
|---|---|---|---|
| Czech Republic | x | | |
| Poland | x | | |
| Saudi Arabia | x | | |
| Turkey | x | | |
| South Korea | | x | |
| Central and Eastern Europe | | x | |
| Middle East | | x | |
| Taiwan | | x | |
| Thailand | | x | |
| Vietnam | | x | |
| China | | | x |
| Indonesia | | | x |
| Italy | | | x |
| Japan | | | x |
| Latin America | | | x |
| Mexico | | | x |
| Peru | | | x |
| Russia | | | x |
| Spain | | | x |
| United Kingdom | | | x |
| Ukraine | | | x |

out of 40 volunteer tourism organisations examined, only 11 requested background checks in terms of CV, references or police and criminal record checks. While there appears to be no data in this regard applicable to the TEFL tourism industry, it is clear that this is an area of concern, where there is the potential for exploitation, whether sexual or otherwise, to be carried out by TEFL teachers who may not have been subjected to the standard background checks that we have come to expect across the western world.

A frequent problem in many destinations is the lack of an institutional framework for the organisational development and management of TEFL tourism. Alongside this, in many countries the presence of several governmental ministries which have intersecting responsibilities regarding the developmental, planning and marketing issues relating to recruiting TEFL teachers and managing TEFL tourism can become a problem since these bodies often have contradicting agendas. While there is the need to balance the growth and development of the tourism industry and to manage the economic and social impacts of this, there is also the need to meet development agendas and manage the education sector in the best way possible.

The need to develop umbrella mechanisms that allow the relevant bodies to work successfully together in order to create the balance necessary for the development of TEFL tourism should be stressed, and it is imperative that public, private and NGOs cooperate in order to achieve any kind of regulation within the TEFL industry. This is, however, easier said than done. The TEFL tourism industry is in its infancy and as such it can be argued that the organisations involved have not yet reached the desired levels of consciousness needed in order to maintain a sustainable and appropriately regulated industry. Barrett's 'model of organisational consciousness' can be used to understand this further (Barrett Value Centre, 2017).

Based on Maslow's hierarchy of needs, Barrett created the 'seven levels model' to provide a clear understanding of human motivations (Fig. 2.4). He argues that all human group structures grow and develop in consciousness in seven well-defined stages. Each stage focuses on a particular existential need which acts as a motivating force in human affairs. The model is frequently utilised to enhance understanding of industry sectors and can be applied to all types of organisations involved in TEFL tourism, ranging from

**Box 2.16.** Sexual exploitation in orphanage tourism

An article on the website Vice (2016) reports on the issue of sexual exploitation in Cambodian orphanages. Vice explains that Cambodia is home to over 500 orphanages – a figure that has doubled in the last decade, resulting in a booming, multi-million dollar industry. Many Cambodian parents think that if they send their children to orphanages they will be looked after and provided for. They will receive an education, access to medical care and better nutrition than they would otherwise have.

This perception has resulted in orphanages no longer being comprised only of orphans, but also children from poor families. In fact, according to a 2012 UNICEF study, an estimated three out of four children in Cambodia's orphanages still have at least one parent. This scheme has now become a loophole for corruption, with some orphanages even offering small sums of money to parents in exchange for their children. There are a large number of unlicensed orphanages in Cambodia that operate without the official documentation required or proper child protection policies. Children in these orphanages are rarely given an education, instead being put to work until the tourists come to visit, where they are used as 'money-making tourist attractions'.

Alongside this, sexual abuse is rife. Western child abusers travel to Cambodia to work in the orphanages in order to gain easy, unsupervised access to the children. In 2007, the Cambodian Children's Trust rescued 14 children from an orphanage where the director had been sexually and physically abusing them. In early 2018, an Australian-run orphanage was closed down due to accusations of child abuse and child trafficking. This orphanage had 'rescued' 21 children from the streets of Phnom Penh and was not legally registered. A week later, a director of another institution in the city of Siem Reap was arrested for sexually abusing two girls, aged 11 and 12. The government promises investigations into such cases but there is rarely evidence of these being carried out.

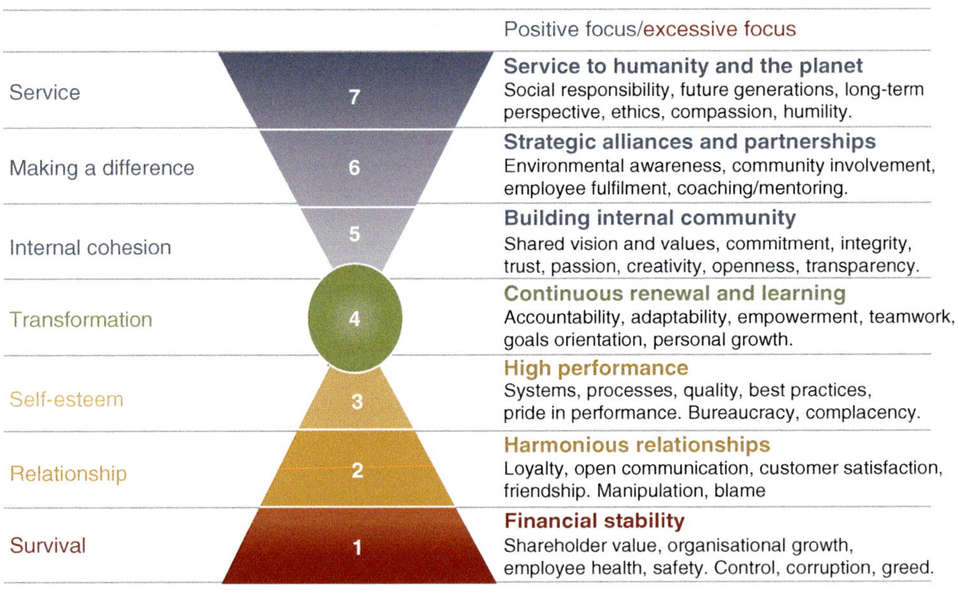


**Fig. 2.4.** Barrett's seven levels of organisational consciousness

large corporations, government departments and municipal agencies to NGOs, development organisations, charities and educational establishments. (Barrett Value Centre, 2017)

Although the model focuses predominantly on the organisation, different levels of consciousness focus on the needs of various stakeholders within the TEFL tourism industry. Fundamentally, company investors and employers are at the core of the first level of consciousness, with employees and TEFL tourists and those working with the education provider being the focus of the second level. The third, fourth and fifth levels continue to focus on the needs of employees and at the sixth level, partners and the local community are introduced into the mix. The seventh level focuses on employees, partners and society in general. The lower needs (levels 1 to 3) focus on the basic needs of business, considering aspects such as financial stability, building employee and customer loyalty, and high performance systems and processes. The focus is then shifted to transformation at the fourth level where there is a shift from fear-based, rigid, authoritarian hierarchies to more open, inclusive, adaptive systems of governance that empower employees to operate with accountability. The higher needs (levels 5 to 7) finally show the need for cultural cohesion and alignment, building and maintaining alliances and partnerships, long-term sustainability and social responsibility. In order to better comprehend the current position within the TEFL tourism industry, and to highlight the extent to which the sector is in its infancy, each level can be further explained in the context of TEFL.

Level 1 focuses on survival. The first need for an organisation is to be financially secure, whether this be achieved through profits, as is the case with commercial enterprises, or through voluntary donations or government funding, as seen with development agencies and charities. Money is perhaps one of the most controversial aspects of TEFL tourism, with the combination of 'supposed' benevolent intentions and for-profit business motivations being somewhat questionable, and it is imperative that this area is managed appropriately in order to satisfy all stakeholders. There is also uncertainty for organisations which are not focused on profit, with reductions in aid funding and philanthropic donations ever-looming.

Level 2 focuses on relationships. Barrett emphasises the importance of harmonious interpersonal relationships and good internal communications, where a sense of loyalty, belonging and caring are integral. Within TEFL tourism there are a number of key players including employees of the TEFL agency, the TEFL tourist and those working within the educational institution involved. It is imperative that all parties have adequate shared communication in order for the business to be successful. This is, however, commonly not the case, with TEFL teachers being allocated unfamiliar responsibilities and educational institutions providing experiences that are not clearly outlined by the TEFL agency. This level of consciousness can be compromised when the TEFL agency and the educational establishment have contradicting views or ideals – for example, if the agency is highly profit-orientated but the establishment is altruistically centred.

Level 3 looks at self-esteem. The focus is on performance and measurement and keeping a balanced and watchful eye on all the key operational indicators, an aspect which may be easier said than done given the diversity of educational institutions and TEFL experiences. At this level, the organisation is focused on optimising its business potential through the use of best practices such as quality, productivity and efficiency. Level 3 organisations tend to be structured hierarchically for the purposes of central control with top-down being the primary mode of decision-making. Rules and regulations are usually developed to bring order to all aspects of their business, although it is important to note that rigid authoritarian bureaucracies risk losing touch with the needs of the local community which is at the heart of the TEFL business.

Level 4 focuses on transformation. Central to this level are the concepts of adaptability, employee empowerment and continuous learning. It is imperative at this level that organisations act on and respond to market advances, paying close attention not only to the TEFL tourism industry in terms of the commodified product and market opportunities, but also to the destination's education sector and wider development agendas. It is important that the voices of both the TEFL agency and the educational institution are heard and that they have relevant input into operational decisions.

Level 5 fosters internal cohesion, building on cultural cohesion and developing a capacity for collective action. The critical issue at this level of consciousness is developing a shared vision of the future and a shared set of values which satisfy both the TEFL agency and the educational system within which it is operating. The key to success is the establishment of a strong, positive, unique cultural identity that differentiates the TEFL organisation from its competitors. The culture of the organisation becomes part of its brand, which in turn becomes more prominent and aligned with the company's ethos and motives.

Level 6 is all about making a difference and deepening the level of internal connectedness within the organisation, as well as expanding the sense of external connectedness outside the organisation. Internally, this can be achieved by helping employees find personal fulfilment in their work. Externally, the focus is on developing partnerships and alliances with businesses, the local community and relevant NGOs. At this level, the organisation can really show its empathy with the local education system and its stakeholders, demonstrating that the organisation's motivations and intentions are aligned with those of the national or regional government and local community.

The final level addresses the concept of service. It is a continuation of the previous level – a further deepening of the levels of internal connectedness, and an expansion of external connectedness. Internally, the focus of the organisation tends to be on producing a climate of ethics, humility and compassion. Externally, the focus will be on local, national or global activism in building a sustainable future for the destination through education. It is at this level that the organisation demonstrates a deep sense of social responsibility and passion towards social justice, human rights, ecology and the global environment. To be successful at level 7, organisations must embrace the highest ethical standards in all their interactions with employees, suppliers, customers, shareholders and the local community. They must always give consideration to the long-term impacts of their decisions and actions.

As the TEFL tourism industry is in its infancy, it can be suggested that the majority of TEFL agencies operating within the market are situated towards the lower end of Barrett's model. This provides some explanation for the lack of regulation of the industry, with an immature perspective of what approaches may be most suitable or which practices may work best. As the industry continues to grow and evolve, issues of best, or worst, practice will become clearer, allowing for regulation to be imposed where appropriate.

There are many models available for use as an informative guide by industry practitioners, such as Barrett's. In an ideal world, there would be the development of such a model specifically for the TEFL tourism industry, although such initiatives rarely work in practice, given the variety of organisations and general complexities. As demonstrated in this chapter, TEFL tourism is no stranger to ambiguity, and given its global scale, spanning a number of different cultural regions, regulation would not be an easy feat. Yet, it seems that some kind of regulation and monitoring is necessary. This will become more apparent as this book progresses and the industry is assessed in further detail.

# 3

# TEFL Tourism as a Form of Niche Tourism

## Situating TEFL Within the Context of Niche Tourism

This book introduces TEFL teaching within the context of tourism, a concept surprisingly unrecognised within the educational or the tourism industry. This chapter seeks to explore this notion through identifying a number of similarities between TEFL tourism and various other forms of tourism, placing it firmly in the 'micro-niche' category. According to Novelli (2005), the author who first articulated the manifestation of a number of alternative tourism forms under the niche tourism umbrella in her book *Niche Tourism: Contemporary Issues, Trends and Cases*, niche tourism is a small and specific tourism type derived from a smaller, broken-down version of the larger market sector (macro-niche) or a specialist small-scale market (micro-niche).

As many authors have proposed, including Novelli (2005) and Yeoman (2008), niche tourism is generally considered to be an alternative, almost an antithesis, to modern-day mass tourism. While mass tourism is inherently homogenous in nature, with standardised products and packages being the focus, niche tourism is defined by its heterogeneous nature, demonstrating elements of individualism and uniqueness. Significant components in the definition of niche tourism include the distinct qualities of niche tourism products, the market size (smaller and more specific than for mass tourism products) and the products containing attractions that appeal to a small, specific group of tourists.

Nowadays, however, it is not as easy as it may seem to polarise tourism forms in this way, but rather many types of tourism fluctuate between mass and niche or may indeed possess elements of both. This was addressed recently in a research note assessing the current state of the volunteer tourism industry. Where volunteer tourism may once have been indisputably identified as a form of micro-niche tourism, Stainton (2016) argues that as a result of the growth of the industry, coupled with its increasing commercialisation and diversity of project types offered, modern-day volunteer tourism should instead be classed as a macro-niche, playing host to an umbrella beneath which there are a range of micro-volunteer tourism forms such as sports coaching, care work, house building, animal conservation and TEFL teaching.

The fragmentation of niches in this way is addressed by Novelli (2005) who describes the way in which niche tourism grows and develops, until it eventually begins to fragment into smaller products and markets. In effect, therefore, the size of the niche grows, but the fact that tourists are participating in it in different ways (through volunteering as a sports coach as opposed to building houses, for example) allows for the different sectors of the niche to be defined accordingly. TEFL tourism is determined as a niche of volunteer tourism and it is important to clarify

**Box 3.1.** Developing niche tourism in the Caribbean

Tourism development consultants in the Caribbean have identified that niche markets are good for diversifying products, minimising the effects of seasonality and reducing pressure on 'honey pot' areas by distributing tourists to lesser-visited places on the islands. There are many tourism forms that are well established in the islands and others that are in development.

The top 20 niche markets are:

- **bird watching**: bird watching facilities, i.e. a reserve or hide, 200 species of bird present;
- **charity tourism**: volunteer, conservation or charity programmes that are available for foreign tourists to participate in;
- **culinary tourism**: food festival of international repute;
- **cultural, heritage and historical tourism**: historical, cultural or heritage attractions;
- **cycling**: bicycle hire and/or cycle tours;
- **diving**: PADI, NAUI, BSAC or CMAS certified dive operator;
- **extreme sports**: bungee jumping, paragliding, kite-boarding, land yachting, hang-gliding, caving, white-water rafting or similar activity;
- **festivals**: home to a national festival;
- **film location**: country has been used as a film location for an international movie;
- **fishing**: facilities for the hire of fishing equipment and/or boats, for freshwater or saltwater fishing;
- **golf**: at least one 18-hole golf course;
- **health, wellness and spa**: spa (or similar) facilities, whether they be a specific spa resort or 'hotel & spa' combination;
- **homestays**: accommodation with a local family;
- **MICE**: facilities for meetings, incentives, conferences, exhibitions for at least 10 delegates;
- **sailing**: facility for yacht charter (bareboat or flotilla) or dinghy hire/instruction;
- **soft adventure**: facilities for adventure/outdoor activities that require little or no experience and are low risk, such as horse riding, snorkelling, canoeing/kayaking and walking;
- **sports tourism**: have held an international spectator-sporting event in the last five years;
- **trekking**: trekking routes that are over five hours long;
- **weddings and honeymoons**: legal and administrative system in place for holidaymakers to get married;
- **wildlife tourism**: unique fauna or products such as turtle nesting grounds, whale/dolphin watching.

(One Caribbean, 2017)

this point. While this may be completely true for those who receive no financial remuneration for their efforts and are therefore classified as 'volunteer' tourists, it does exclude a significant proportion of the TEFL community who are rewarded with a salary.

It is clear, therefore, that TEFL tourism cannot be strictly identified as a micro form of volunteer tourism. In fact, there appears to be no clear 'macro umbrella' beneath which TEFL tourism neatly fits. Rather, there are a number of tourism forms within which TEFL tourism manifests, demonstrating links between tourism forms instead of a macro tourism form/singular body of literature from which the concept of TEFL tourism derives, in accordance with Novelli's (2005) traditional macro-micro model (Fig. 3.1). Preliminary studies addressing the concept of TEFL tourism (Stainton, 2017b, c; 2018b) identify links with a number of macro-niche tourism forms including: volunteer, philanthropic, cultural, pro-poor, educational, adventure and gap year or backpacker travel. It is also clear that within the concept of TEFL tourism, notions of teaching, linguistics and service learning make up an inherent part of the experience.

It can be argued that niche tourism is growing in significance because of a shift in motivational factors for travellers, where tourists have developed as consumers. Alongside the increasing desire among members of western society to 'do something good' and to enhance their CVs, it is no surprise that there has been an evident convergence of the two concepts of tourism and

**Box 3.2.** Overview of volunteer tourism project groups and activities

| Group | Activity | Number of projects |
|---|---|---|
| Community welfare | Community development | 805 |
| | Childcare | |
| | Elderly | |
| | Disability | |
| | Peace | |
| | Human rights/legal | |
| Teaching | Teaching a foreign language (TEFL) | 572 |
| | Sports coaching | |
| Environmental | Nature conservation | 502 |
| | Wildlife protection | |
| | Global warming | |
| Medical | Hospital support | 236 |
| | Pandemic support (HIV, Ebola) | |
| | Drug rehabilitation | |
| Cultural | Arts | 131 |
| | Cultural exchange | |
| | Women empowerment | |
| | Heritage conservation | |
| | Museum support | |
| Business development | IT support | 91 |
| | Accounting support | |
| | Farming/organic | |
| Building | Construction | 54 |
| | Renovation | |
| | Water management | |
| Research | Wildlife monitoring | 47 |
| | Land mapping/zoning | |
| Other | Catastrophe relief | 28 |
| | Olympic games | |

(Tomazos and Butler, 2009)

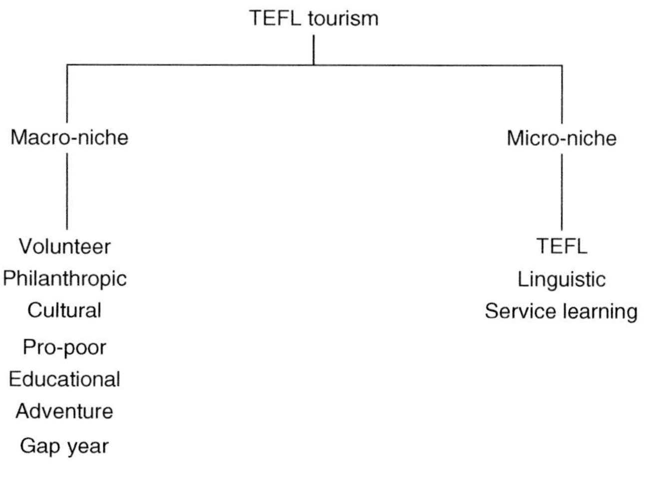


**Fig. 3.1.** TEFL tourism links with micro and macro tourism forms

teaching. While this book recognises the distinct lack of research focusing on TEFL tourists as consumers due to the infancy in the current body of literature, it is important to note that there is a broad body of knowledge within the wider tourism-based literature upon which TEFL tourism researchers can draw.

Cohen (2014), in his review of consumer behaviour studies in tourism, concludes that tourism researchers have historically reviewed individual concepts (e.g. motivations, perceptions, loyalty), specific influences (e.g. technology, generation Y) and particular research contexts (e.g. under-researched segments, individual versus group decision-making). While such literature can be useful in attempting to understand the TEFL tourism marketplace, it is neither prescriptive nor specific. In fact, Cohen concludes that the existing body of research on travel behaviour is fragmented due to: individual studies replicating one or a few consumer behaviour concepts borrowed from marketing and general management fields and applying them to tourism; research contexts disabling the ability for generalisation; an overuse of quantitative approaches; and few studies adopting the use of longitudinal and/or holistic approaches to understand the behaviour or processes being investigated.

The limitations highlighted here affect not only the broader body of tourism-based literature, but also the niche tourism market and, more particularly, TEFL tourism. In effect, Cohen (2014) highlights a shortfall in the understanding of specific consumers who, in the end, are the drivers of the tourism industry. Niche tourism has traditionally been hailed as being a more sustainable approach to tourism development than its mass tourism counterpart. However, this is a difficult argument to make if there is not a true and thorough understanding of the consumers and their behaviour. While the intention of this chapter is not to discuss the sustainability of niche tourism, nor to analyse the many theories addressing consumer behaviour, the shortfalls in literature identified do help to emphasise the point that there is a lack of understanding of tourists and their behaviour, both within TEFL and niche tourism and the wider tourism community. In order to ensure sustainable management of TEFL tourism, it is imperative to understand the behaviours and decisions of consumers as well as any impacts, positive and negative, that TEFL tourism may have. To begin to assess these aspects, it is important to identify with relevant tourism forms and associated literature and to learn lessons from these, both the good and the bad.

## The Work-leisure Binary

Traditionally, the general notion of the tourist role is separated from work-based activities – the tourist is defined by the tendency to spend money rather than to work for it. TEFL tourism is unique, however, in that it involves varying degrees of both work and leisure as core parts of the tourist experience. It offers the tourist two, often contradictory, concepts: work and leisure. Perhaps one of the reasons for the lack of convergence to date between the two concepts of TEFL teaching and tourism is the inherent incongruity of leisure and work as fields of human activity. Several academics have supported the notion that throughout history tourism has been the antithesis of work. Oglivie claimed, back in 1934, that the tourist should be characterised not as a producer but as a consumer. More recently Graburn (1989) proposed that 'our conception of tourism is that it is *not work*' (p. 22), while Urry (1990) suggested that the places gazed upon ' ... are for purposes which are not directly connected with paid work, and normally they offer some distinctive contrasts with both paid and unpaid work' (p. 3). With the abundant growth of niche tourism forms in modern-day society, however, such polarisation is no longer valid, with the lines between work and leisure becoming increasingly blurred.

There are now many tourism forms within which there is a core interaction between the concepts of work and leisure. Volunteer tourism's many micro-niches fit this bill perfectly, with activities such as sports coaching, house building, providing medical assistance or undertaking biodiversity research all requiring a substantial 'work' element as part of the experience. There are also paid tourist opportunities that have recently been introduced into the proverbial mix such as business tourism, WOOFING (worldwide opportunities on organic farms) and TEFL tourism. Despite the increasing recognition of tourism experiences containing work-based elements, there remains only a small body of academic literature addressing this.

Means of examining the inherent contradiction of work and leisure have largely focused around the concept of working holidays (Cohen, 1974), the roles played by obligation (Stebbins, 2000; Lepp, 2008), and serious leisure (Stebbins, 2000). The term 'working holiday', under-researched, is not a new concept. It was coined by Cohen in 1973, who defined working tourists as those who travel from one country to another to work for short periods of time (p. 91). Uriely (2001) then developed a typology of working tourists in 2001, comprising four working tourist types (Fig. 3.2). The *travelling professional worker* is typically characterised as a middle-class, mature tourist whose travel is predominantly for business purposes. The *migrant tourism worker* is motivated to travel for the dual purpose of both securing work in a tourist hub but also to enjoy a holiday experience. The *non-institutionalised working tourist* is a tourist who engages in unskilled and low-level, repetitive manual labour occurring in various fields in order to finance their travel, and the *working holiday tourist* enjoys working or volunteering as part of the tourist experience. While there is an inevitable amount of overlap between the types of working tourist identified by Uriely, it is likely that a TEFL tourist will fall within one of the three latter working tourist types. It is also suggested that there is a variable missing in Uriely's analysis – the *career-focused tourist*.

The career worker is one who wishes to gain relevant work experience from their TEFL placement, whether paid or unpaid. The intention is to boost their CV and for their TEFL experience overseas to help them climb the career ladder for which they aspire. The migrant worker is one who secures a position as a TEFL teacher in order to reside in the destination within which their job is based. This is particularly relevant for destinations which may appeal to tourists due to aspects such as climate, cost of living, culture, leisure activities or language. The non-institutional tourist (Box 3.3) is one who is unskilled and is working simply to maintain their desired lifestyle, such as living in a particular destination or travelling. This tourist tends not to enjoy their job, but rather they do it as a 'means to an end'.

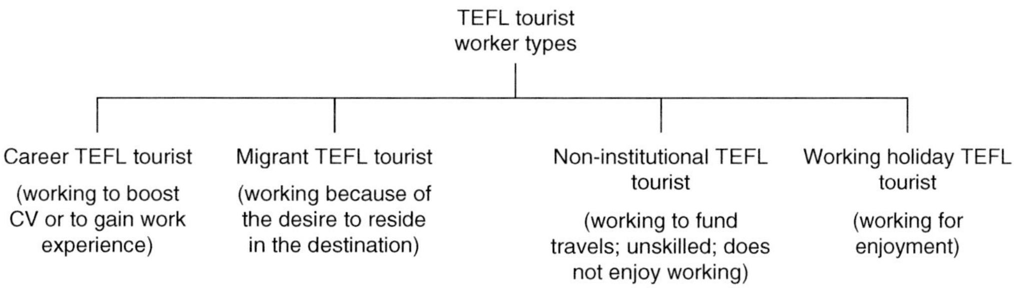


**Fig. 3.2.** TEFL tourist worker types

**Box 3.3.** The non-institutional TEFL tourist

While for some, notions of altruism and goodwill may be the primary motivation for undertaking a teaching placement, this is not the case for all TEFL teachers. Appealing tourist destinations frequently attract travellers who are hedonistically motivated, prioritising their enjoyment above their teaching responsibilities. Home to the infamous 'full moon party', buckets full of alcohol and an array of nightlife opportunities, Thailand welcomes party-goers into its classrooms daily. While nursing a hangover or, at best, surviving on little sleep, these tourists are paid to 'educate' classes of 50+ Thai children. It is common for said tourists to take sick days from work in exchange for travel opportunities or rest. Furthermore, it has been known for TEFL tourists such as these to 'buy' their TEFL counterfeit certification from street vendors selling illegal fake documentation. These TEFL tourists do not centre their focus around their teaching practice, and often dislike teaching. They do it only to prolong their lifestyle of parties and touristic activities.

Lastly, the working holiday TEFL tourist is a teacher who works for enjoyment alongside a variety of tourist-based activities.

There are three dominant ways in which one can undertake work within the 'holiday' sphere. Some working holidays take the form of a volunteer programme, in which the working activity is unpaid. There are a number of tour operators who have capitalised on the popularity of this type of holiday recently, often charging high prices to the tourist to undertake such a trip. This is generally known as volunteer tourism. Other holidays are designed for regular tourists who pay for an experience that includes optional work. While this is not yet as developed within the tour operator market as volunteer holidays, operators are beginning to buy into such programmes with the likes of WOOFING and TEFL tourism becoming more common in the marketplace. Lastly, there are dynamic working holidays, where the tourist organises the elements of their trip independently. This is particularly common in countries which offer working holiday visas, such as the UK and Australia. In the TEFL tourism industry this is a popular way of securing employment upon arrival overseas.

In terms of the work-leisure binary, it is evident that each concept plays a stronger or weaker role within the different TEFL tourist worker types. It is likely, for example, that a career TEFL tourist will focus more strongly on the work element of the experience than the holiday element. In contrast, the working holiday TEFL tourist may place heavier emphasis on the leisure aspects. In fact, it can be argued that working holiday TEFL tourists exemplify the most tourist-oriented manifestation of the interaction between work and tourism.

Rather than assessing work and leisure in a binary fashion in this way, however, Stebbins introduced the concept of 'serious leisure' in the late 1970s amid increasing notions of individualism and the related leisure trends of commodification and consumption. Serious leisure is defined as: 'the systematic pursuit of an amateur, hobbyist, or volunteer core activity that people find so substantial, interesting, and fulfilling that, in the typical case, they launch themselves on a (leisure) career centred on acquiring and expressing a combination of its special skills, knowledge and experience' (Stebbins, 1992, p. 3). Serious leisure participation is typically identifiable by six qualities: the occasional need to persevere; the acquisition of specialised knowledge, training, experience or skill; the ability to follow a career path in the activity; the adoption of a common ethos of values, attitudes and practices; strong identification with the serious leisure pursuit; and the experience of durable personal and social benefits (Stebbins, 1992).

In 2007, Stebbins introduced the 'serious leisure perspective' (SLP), a theoretical framework that bridges and synthesises three main forms of leisure, known as serious leisure, casual leisure and project-based leisure. Within each form, a variety of types and subtypes have also emerged over the years, as pictorially summarised by The Serious Leisure Perspective website (Fig. 3.3).

According to Stebbins (1997), casual leisure is immediately, intrinsically rewarding. It tends to be a short-lived, pleasurable activity requiring little or no special training to enjoy it. It is fundamentally hedonistic and is engaged with for reasons of pure enjoyment or pleasure. Casual leisure typically includes aspects such as play, relaxation, passive entertainment (e.g. TV, books, recorded music), active entertainment (e.g. visiting tourist attractions, playing games), sociable conversation and sensory stimulation (e.g. sex, eating, drinking). Casual leisure makes up an integral part of the TEFL tourism experience, with tourist-based elements, relaxation, etc. being important parts, regardless of whether the tourist is leisure or work orientated. It can also contribute to the classroom experience with elements such as playing games or listening to music being common teaching practices within the EFL classroom.

At a halfway point between casual and serious leisure is what Stebbins (2005) describes as project-based leisure. This consists of short-term, moderately complicated activities which can be either one-off or occasional undertakings. Such leisure typically involves considerable planning, effort, and sometimes skill or knowledge. Project-based TEFL placements are common, with short-term placements widely available along with ad hoc work such as English camps or private tuition.

Serious pursuits is the umbrella concept encompassing serious leisure and devotee work and is likely to include TEFL tourists who are

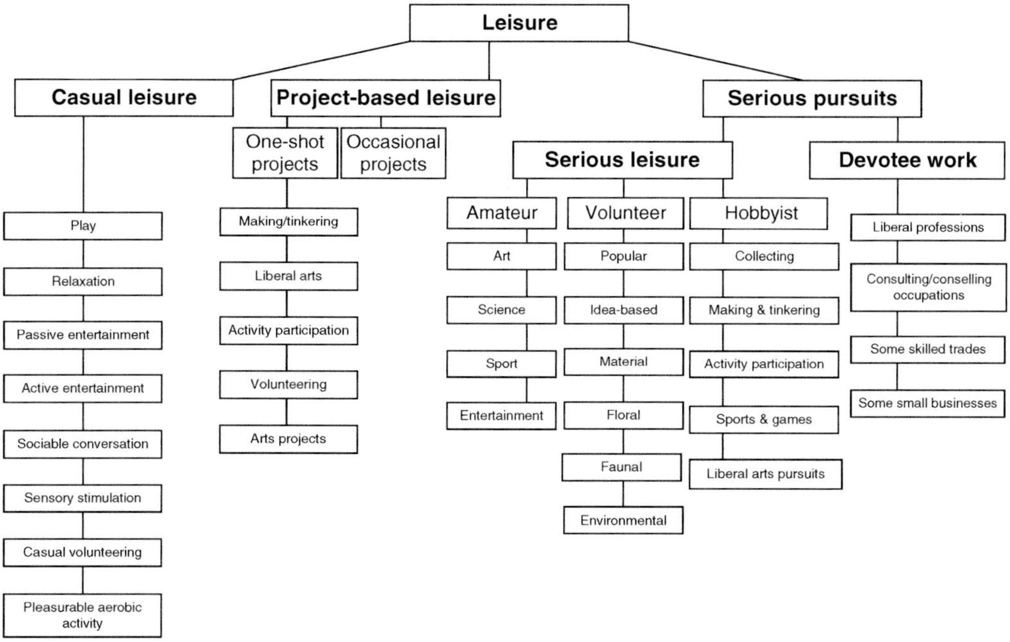


**Fig. 3.3.** The serious leisure perspective (The Serious Leisure Perspective, www.seriousleisure.net/diagrams, 2013)

predominantly altruistic or career-minded. In accordance with Stebbins' (2007) categorisation of amateur, volunteer and hobbyist tourists, TEFL tourists do not necessarily have to be qualified or highly skilled to be classed as a serious tourist; rather they need to demonstrate a willingness and commitment for the employment which they are undertaking. There is, of course, some overlap between the types of serious leisure participation, and indeed, an individual engaged in serious leisure might take on more than one of these roles through their involvement.

## Volunteer Tourism as a Macro

As previously noted, TEFL tourism has close ties with volunteer tourism, the main difference being the presence of financial remuneration or incentives. In fact, referring to the body of literature addressing the volunteer tourism industry is particularly useful in beginning to attempt to comprehend the TEFL tourism industry, due to the attention paid to this field, both by academics and practitioners, over the past two decades. Increasing scrutiny and critical assessment of volunteer agencies and operations has brought to light a number of areas warranting careful attention and, given the similarities between the TEFL and volunteer tourism sectors, there are many lessons that can be learned from this macro-niche and applied to the micro-niche of TEFL tourism.

Volunteer tourism, also known as 'voluntourism' or a volunteer vacation, is often discursively constructed as it overlaps a number of research areas. Associations have been made between the fields of tourism research and volunteering and leisure studies, in addition to a variety of forms of tourism (Lyons and Wearing, 2012; Mostafanezhad, 2013). Fundamentally, the concept of volunteer tourism comprises both volunteerism and tourism and can today be found in virtually every sector, even in the extremes of mass tourism (Butcher, 2011; Wearing and McGehee, 2013). In line with the general perspective taken towards niche tourism, volunteer tourism plays on the pejorative connotations that have accompanied the evolution of mass and package tourism and their negative impacts, which frequently relate to environmental degradation and sociocultural disturbance (Robinson and Novelli, 2007). It is this sociocultural aspect that

**Box 3.4.** i-to-i Cambodia TEFL experience itinerary

Day 1
**Arrival in Cambodia**
Make sure you arrive at Phnom Penh airport between 6am and 8pm, our team will meet you there to transfer you to your hotel. You'll get the chance to meet the other interns and settle in over a welcome meal.

Day 2–3
**Orientation in Phnom Penh**
During your 2 day orientation you'll learn about Khmer culture and history, what your role at your school will be like, and you'll even get some language lessons!

Day 4–84
**Your teaching experience**
For 12 action packed weeks, you'll be teaching at a private school within Phnom Penh, with students aged between 3 and 16. You'll be living on the school campus, sharing a room with at least one other intern, with Wi-Fi available and lunch provided by the school during teaching.

Weekends are yours to enjoy, and interns usually leave their school campus to explore Cambodia. There are loads of fascinating things to do and see, plus our in-country staff are on hand with expert suggestions on how to spend your spare time.

Day 85
**Transfer to Phnom Penh**
Today you'll be transferred back to central Phnom Penh for a night catching up with the other interns before your flight home, onward travel or the optional trip to the lively resort of Sihanoukville the next day.

Day 86–91
**Beach trip to Sihanoukville (optional)**
Now time for 5 days' pure relaxation, at 'one of the most beautiful bays in the world'! In lively Sihanoukville there are loads of gorgeous beaches, pumping bars, shops and food outlets plus exciting boat trips to take with your fellow interns in between catching up and reminiscing.

Day 92
**Time to say goodbye!**
You'll then be transferred back to Phnom Penh for a final special night eating, drinking and saying goodbye. Help will be on hand from our team to organise onward travel or the journey home.

(i-to-i, 2017)

encompasses the notion of TEFL teaching. According to Wearing (2001, p. 1) '[volunteer tourists are] those tourists who, for various reasons, volunteer in an organised way to undertake holidays that might involve aiding or alleviating the material poverty of some groups in society, the restoration of certain environments or research into aspects of society or environment'.

Although frequently used throughout the literature on volunteer tourism, it can be argued that this definition is no longer contemporary, since it fails to address the expansion of volunteer tourism, its manifestation in a range of contexts and its progressive movement away from being a micro- to a macro-niche (Wearing and McGehee, 2013; Stainton, 2016). It has been suggested that, as a result of the continually growing and evolving industry, scholars should take a more specific approach, analysing the different types of volunteering individually in order to facilitate in-depth analysis that is specific to that particular activity (Stainton, 2016). One such example is the study of TEFL teaching. This phenomenon may, in some instances, fall under volunteer tourism, whereas in others it may not (Stainton, 2016). This is particularly relevant to paid TEFL teaching. Traditionally, volunteer tourism has been associated with those who receive no monetary recompense for their efforts (Brudney, 2000; Wearing, 2001; Ellis, 2003). However, the blurring of paid and voluntary work has become commonplace (Lyons, 2003;

Lyons and Wearing, 2012), with many organisations providing incentives such as reimbursements, cash and qualifications as part of the volunteering experience. Conversely, Tomazos and Butler (2009) claim that volunteer tourism projects are not 'true' opportunities if they are internships or paid employment. While some researchers may thus classify a teaching placement as a form of volunteer tourism, others may not (Stainton, 2016).

The last two decades have seen a steady increase in the scholarship of volunteer tourism, with the phenomenon moving from the periphery of tourism research towards the centre, aligning more closely with macro forms of niche tourism and in some instances demonstrating distinct similarities with the mass tourism sector. Volunteer tourism has experienced exponential growth in recent years and the latest available statistics valued the industry at £77.2 million and volunteer tourist spend at £832m–£1.3 billion per year (Tourism Research and Marketing (TRAM) (2008). Similar to TEFL experiences commodified and sold by agencies (Stainton, 2018a), volunteer tourists typically pay for the privilege of volunteering. This payment frequently covers housing, meals, project materials, administration and on-site staff support (Tomazos and Butler, 2009; Keese, 2011).

Volunteer destinations range from local to long-haul (Brown, 2005), and the average volunteer trip costs the participant approximately £2,000 (TRAM, 2008). With the prolific commercial growth of the industry and developments in the world economy, it is important to note that these costs are likely to have increased since the time of publication. The average volunteer tourist is aged 18–25 years (Wearing, 2001; Broad and Jenkins, 2008; TRAM, 2008; Tomazos and Butler, 2009), the majority are female (Brown, 2005; TRAM, 2008), the majority are single (Gecko *et al.*, 2009; Tomazos and Butler, 2009) and volunteer tourists tend to be of a white ethnicity (Birdwell, 2011). Volunteer tourists tend to be educated or undergoing education to university level (Brown, 2005; Tomazos and

**Box 3.5.** China volunteer teaching project

Below is an advert from GapYear.com (2018).

**Duration: From 2 weeks**

**Cost: From £595**

From kindergarten to college, the Frontier-China teaching project offers immense help to the impoverished children who wish to learn English. Following the local school timetables, you will be at the school five days a week, teaching between two and four hours a day. The evenings and weekends are yours to spend as you like, and the local area offers a variety of activities for all tastes.

**Pre-departure inclusions**

Pre-departure support
Travel & medical advice & documentation
Equipment advice
Discounted medical kit
Free Frontier T-shirt
UK TEFL training day & certification

**In-country inclusions**

Food during the week
Accommodation
Local orientation
Project training
Ground transfers & in-transit accommodation
Airport pick-up
In-country emergency support
24-hour international HQ backup

Butler, 2009; Birdwell, 2011) and most tend to be continuing students or those undertaking a gap year (Gecko *et al.*, 2009), with the average placement lasting two to four weeks (Callanan and Thomas, 2005; Gecko *et al.*, 2009; Keese, 2011).

While there is a marked lack of literature pertaining to such characteristics within the TEFL tourism industry, preliminary research based on TEFL tourists in Thailand does indicate that there are similarities between TEFL tourists and their volunteer colleagues (Stainton, 2018b). The majority of TEFL tourists were single and 85% were qualified to degree level or above. TEFL tourists did differ in terms of age, however, as they tended to be older than their volunteer tourist colleagues, with the majority (58%) being between the ages of 22–30.

Initial research into the volunteer tourism phenomenon largely focused around motivations benefiting the host community such as altruism, giving something back and helping others (Wearing, 2001; Brown, 2005; Coghlan, 2006; Soderman and Snead, 2008; Gecko *et al.*, 2009; Leonard and Onyx, 2009; Sin, 2009; Chen and Chen, 2011; Lo and Lee, 2011; Wickens, 2011). Studies have, however, also demonstrated the significance of a number of motivational factors beneficial to the volunteer tourist such as developing skills and experience to aid career progression (Wearing, 2001; Coghlan, 2006; Soderman and Snead, 2008; Coghlan and Fennell, 2009; Daldeniz and Hampton, 2011), personal development (Coghlan, 2006; Soderman and Snead, 2008; Benson and Seibert, 2009; Gecko *et al.*, 2009; Wickens, 2011), cultural immersion (Brown, 2005; Coghlan, 2006; Benson and Seibert, 2009; Gecko *et al.*, 2009; Sin, 2009; Chen and Chen, 2011; Lo and Lee, 2011; Wickens, 2011; Taillon and Jamal, 2013) and the desire to do 'something meaningful' (Coghlan, 2006; Gecko et al., 2009).

However, as research into volunteer tourism has progressed, academics have begun to express the need for a more cautionary approach, highlighting that not all the motivations previously identified are primary reasons for participating in volunteer tourism projects (Coghlan, 2006; Broad and Jenkins, 2008; Soderman and Snead, 2008; Benson and Seibert, 2009; Sin, 2009; Chen and Chen, 2011; Daldeniz and Hampton, 2011; Wickens, 2011). A number of external influences may be important factors. These include, for example, the desire to learn the local language (Soderman and Snead, 2008) or to live in another country (Benson and Seibert, 2009; Wickens, 2011); the influence of family or peers (Soderman and Snead, 2008; Chen and Chen, 2011), religion (Brown, 2005; Taillon and Jamal, 2013) and cost (Coghlan and Fennell, 2009; Gecko *et al.*, 2009; Sin, 2009; Chen and Chen, 2011). The reputation of the volunteer tourism organiser sometimes plays a role in motivation (Lo and Lee, 2011) along with the structure of the programme offered (Wearing, 2001; Soderman and Snead, 2008) and whether the placement is part of a course requirement (Lyons, 2003; Lyons and Wearing, 2008; Daldeniz and Hampton, 2011). Education is a frequently noted motivation (Wearing, 2001; Brown, 2005; Coghlan, 2006; Benson and Seibert, 2009; Gecko *et al.*, 2009; Leonard and Onyx, 2009; Lo and Lee, 2011). There are additionally some motivations, such as sex tourism, which have seldom been researched (Wright, 2014).

Once again, initial research into the motivations exhibited by TEFL tourists demonstrates several similarities with many of the studies named above (Stainton, 2018a). Key motivational themes for becoming a TEFL tourist can be broadly organised into aspects relating to travel and tourism, life experiences such as skill development or new opportunities, and destination-specific motivations such as Thai food or culture. Over half of the TEFL tourists in Thailand examined by Stainton (2018a) strongly agreed that travel and tourism was a motivating factor, thus emphasising the links between TEFL and tourism, and for some, TEFL teaching was secondary to the travel experience. Motivations exhibited by TEFL tourists in this study included having the opportunity to visit the beach and to become involved with the nightlife activities and having the opportunity to immerse themselves into the local culture – experiencing cultural aspects such as religion, ethnic traditions, linguistics and learning about Thai heritage through the community engagement. Motivations thematically classified as life experiences (defined as any endeavour which facilitates experiences that the TEFL teacher may not otherwise have had), included a desire to enhance their CV, skill development and undertaking a trial of teaching as a career. Some TEFL tourists demonstrated that they were motivated to learn more about themselves. The majority in this study identified that they were motivated by prospects of 'doing something good'. However, as suggested by Lyons and Wearing in their study of volunteer tourism (2012), this may not necessarily be indicative of their primary motivations. Motivations such as the reduced cost of living, a slower pace of life and the ability to easily secure employment can be attributed to destination-specific motivations.

Early research suggested that volunteers would fit into one of two motivational categories: those who are primarily volunteer-minded and those who are primarily vacation-minded (Brown and Morrison, 2003). Volunteer-minded individuals tend to devote the majority of their time to volunteer activities at the destination (Brown, 2005), emphasising their altruistic motivations. The antithesis of this type is the vacation-minded tourist, who is likely to spend only a short amount of time volunteering during their trip and a considerable amount of time pursuing the hedonistic elements (Brown, 2005). Similarly, Daldeniz and Hampton's (2011) typology classes volunteer tourists as VOLUNtourists and volunTOURISTS. Here the VOLUNtourist is one that is predominantly motivated by prospects of helping those less fortunate, whereas the volunTOURIST is primarily motivated by their desire to travel and explore. While a VOLUNtourist would immerse themselves within the local culture, a volunTOURIST would see themselves as an outsider or a traveller. This is a differentiation that can also be applied to TEFL tourists, categorised according to whether the tourist is primarily motivated by teaching or by tourism.

A more comprehensive typology of volunteer tourists is that of Callanan and Thomas (2005), who propose that volunteer tourists are best categorised by motivation based on their choice of destination, duration of project, focus on experience (self-interest versus altruism), qualifications, active versus passive participation and level of contribution to local people. 'Shallow' volunteers are classified as those who are dominated by personal interest, preferring short trips (normally a few weeks), demonstrating few skills, experiencing low levels of direct contribution to locals and tending to be more passive in participation (Chen and Chen, 2011), consequently impacting on their ability to contribute significantly to the local environment or community. These volunteer tourists are focused on self-interest and self-development and demonstrate motives such as CV and ego enhancement (Holmes and Smith, 2012). At the other end of the continuum, those volunteers who tend to have greater consideration for the community are classified as 'deep' tourists. These tourists typically participate in projects that are longer in duration (at least six months) and possess specific skills and qualifications. For these tourists, the project and the impact on the environment and community are key to the volunteer experience.

Volunteers are driven by a mix of motivations and motivation factors are complex and multifaceted (Wearing, 2004). Soderman and Snead (2008) and McGehee *et al.* (2009) separate volunteer tourists into three categories: the *vanguards*, the *pragmatists* and the *questers*. The most motivated group are the vanguards, who demonstrate interest in aspects such as skills-building and seek the most physically and mentally intense volunteer tourism experiences. This is the smallest of the three categories and tends to comprise the youngest participants. The largest group are the pragmatists. These tend to be middle-aged volunteers that are motivated by the prospects of developing relationships with locals and demonstrate a desire for human connection. The third group, the questers, tend to be the oldest and demonstrate a lack of awareness of their motivations. However, in contrast to the vanguards, who tend to demonstrate self-interested motivations, they gravitate towards the altruistic (McGehee *et al.*, 2009).

The volunteer tourist has long been a challenge to define (Gecko *et al.*, 2009) and the lack of clarity about the role of a volunteer tourist has created what has been described as 'role ambiguity' (Lyons and Wearing, 2008). Many volunteers do not perceive themselves as tourists (Wearing, 2001; Stoddart and Rogerson, 2004; Lepp, 2008) and some actively promote an anti-tourist theme (Lepp, 2008). By contrast, some volunteers perceive themselves as a tourist at various points during their trips (e.g. when they have completed their volunteering duties) and as a volunteer at others (Sin, 2009). Mustonen (2005) similarly identifies the shifting roles of volunteer tourists, who may go from being tourists seeking

**Box 3.6.** BUNAC's (2018) top five reasons to do a TEFL course and teach English abroad

**1. Do something worthwhile on your travels**

Employment for young people in countries such as Chile and South Africa relies on their ability to speak English. Rather than just travelling, why not take the opportunity to help these young people have a better future!

**2. Learn a new language**

Similarly, with the modern world becoming more and more integrated, being bilingual has never been more important – there is no better way to learn a new language such as Spanish or Chinese than by living there!

**3. Learn some new skills**

Teaching English abroad not only gives you teaching skills but also improves your confidence in public speaking, your communication skills, your patience and your ability to be flexible in situations that demand a change of tack.

**4. Discover a new culture**

In the majority of volunteer abroad schemes you will be living with a host family and as such be submerged in their ways of life – there is no better way to learn to love the simplicities of life such as a warm shower than living in under-developed parts of the world.

**5. Meet like-minded people**

Volunteering abroad only appeals to people seeking unique experiences and as such you will meet some of the most like-minded similar people on your trip abroad!

pleasure, relaxation and stimulation during part of their trip, to altruistically helping the community or environment at other times.

Proponents of typological research argue that this provides a valuable foundation for understanding and acting within the respective market (Lyons and Wearing, 2008). However, it can be argued that the models discussed are simplistic since they do not take the different segments of the volunteer tourism industry, such as the TEFL sector, into account (Stainton, 2016). Furthermore, the way in which tourists are classified is largely based on opinion as opposed to systematic evidence. It is also worth noting that the existing research surrounding volunteer tourism motivation is not entirely representative. Firstly, the industry is now emerging into the Asian and African markets (Lo and Lee, 2011; Alexander, 2012), so the majority of existing research does not account for tourists from these backgrounds. Secondly, research into the motivations of volunteer tourists within a particular sector of the industry, such as TEFL, is limited, thus questioning the value of research taking a broad approach (Stainton, 2016). Despite such limitations, typologising tourists in this way can be a worthwhile endeavour when attempting to understand tourist types, their motivations, behaviours and characteristics, particularly when little is known about said tourists. Given this conclusion, a typology of TEFL tourists has been proposed – drawing upon the literature discussed above – later in this book.

## Cultural TEFL Tourism

Given the inherent cultural interaction between TEFL tourists and the society within which they are working, cultural tourism makes up an integral part of the TEFL tourism experience. TEFL teaching provides access to the local culture in two spheres, firstly through undertaking excursions to cultural attractions such as religious sites, historical monuments or museums, and secondly through community engagement while working in an educational establishment. According to the World Tourism Organisation (WTO) (2000), cultural tourism can be broadly defined as the movements of persons who satisfy the human need for diversity, tending to raise the cultural level of the individual and giving rise to new knowledge, experience and encounters. Cultural tourism is commonly associated with education in this respect, and some researchers describe it more narrowly as 'educational cultural tourism' (e.g. Richards, 2005; Buhalis and Costa, 2006; Horner and Swarbrooke, 2007). The inherent link between cultural tourism and education emphasises the relevance of this form of tourism to the examination of TEFL tourism, with education being at the core of the experience.

Although a common, more specific definition has not been agreed among academics due to the complexity and subjectivity of the term 'cultural tourism', there do appear to be two distinct viewpoints. The first focuses on the consumption of cultural products such as sites or monuments (Bonink, 1992; Munsters, 1994). The second comprises all aspects of travel, where travellers learn about the history and heritage of others or about their contemporary ways of life or thought (MacIntosh and Goeldner, 1986). Csapo (2012) contends that the umbrella term 'cultural tourism' can encompass a number of tourism forms, including heritage (material – e.g. historic buildings and non-material – e.g. literature, arts); cultural thematic routes (e.g. spiritual, gastronomic, linguistic); cultural city tourism; traditions/ethnic tourism; events and festivals; religious tourism; and creative culture (e.g. performing arts, crafts). While TEFL teaching can be argued to constitute a cultural thematic route in its own right as a result of inherent interactions with the local community and its culture, the TEFL experience also integrates other aspects of cultural tourism identified by Csapo (2012).

In an attempt to understand the scope of cultural tourism within the TEFL experience, relevant typologies can be examined, which are generally based on the tourist's level of motivation. Bywater (1993), for example, differentiated tourists according to whether they were culturally interested, motivated or inspired. *Culturally interested* tourists demonstrate a general interest in culture and consume cultural attractions casually as part of a holiday rather than consciously planning to do so. *Culturally motivated* tourists consume culture as a major part of their trip, but do not choose their destination on the basis of specific cultural experiences whereas, for *culturally inspired* tourists, culture is the main goal of their holiday.

A more complex typology was proposed by McKercher and Du Cros (2002), who grouped tourists based on the depth of the cultural experience sought, separating them into one of five hierarchical categories. The first is the *purposeful cultural tourist* for whom cultural tourism is their primary motive for travel. These tourists have a very deep cultural experience. The second category comprises the *sightseeing cultural tourist* for whom cultural tourism is a primary reason for visiting a destination, but the experience is more shallow in nature. Thirdly the *serendipitous cultural tourist* does not travel for cultural reasons but, after participating in cultural activities, ends up having a deep cultural tourism experience. The fourth category, the *casual cultural tourist*, is weakly motivated by culture and subsequently has a shallow experience. Lastly, the *incidental cultural tourist* is one who does not travel for cultural tourism reasons, but nonetheless participates in some cultural activities and has shallow experiences.

Adapting this typology, Petroman *et al.* (2013) segment tourists based on their preferred cultural activities. The purposeful cultural tourist, identified by McKercher and Du Cros (2002), enjoys learning experiences that challenge them intellectually and visits history museums, art galleries, temples and heritage sites that are less well known. The *tour amateur cultural tourist* is akin to the sightseeing cultural tourist of McKercher and Du Cros. These tourists often travel long distances, visit remote areas and enjoy tours and wandering through the streets. For the *occasional cultural tourist*, culture plays a moderate role in the travel decision. This tourist enjoys an insignificant cultural experience, their preferred activities being to visit attractions and temples that are easy to reach and explore, although not to the extent that the tour amateur cultural tourist does. For the incidental cultural tourist, culture plays a small or no role in the decision to travel. They enjoy an insignificant cultural experience while visiting attractions that are within easy reach or are heritage theme parks. Petroman *et al.*'s (2013) final segment is the *accidental cultural tourist*. Culture plays a small or no role in the decision to travel, yet the accidental cultural tourist enjoys a deep cultural experience. This tourist type is diverse and as such has no preferred activities attributed to it. While these typologies are somewhat vague and subjective in nature, they can be utilised as a means to understanding the extent to which cultural tourism plays a key role in the TEFL experience.

**Box 3.7.** Life as an English teacher in Japan

The following is an excerpt from the LSE blog (2018), written by Francesca de Munnich, MSc alumni, LSE.

> Moving to Japan to teach was incredibly challenging at first for a number of reasons: not being able to speak the language, being so far from home and living and working in a completely different and unique culture. Yet these turned out to be the aspects that made teaching English abroad such a fascinating experience. Once I arrived, I was warmly welcomed by my students, colleagues and neighbours alike and quickly settled into my new life and job. I taught full time at a high school and also visited a special needs school once a month. Both experiences were demanding yet rewarding.
>
> I have many cherished memories from my year in Japan, both inside and outside work. Highlights included climbing to the summit of Mt Fuji, watching sumo wrestling and geisha performances, as well as experiencing the myriad of weird and wonderful festivals. What was perhaps most memorable, however, was seeing my students grow in confidence and enjoy using the English language. I believe there is nothing more powerful than finding a way to connect with others and I hope that my former students will be inspired to use English, not just in their studies, but also beyond the classroom to communicate and see the world.

## Philanthropic TEFL Tourism

Although there are a magnitude of motivations for taking part in TEFL tourism, for some, philanthropy is an important part of the experience. Within the tourism industry, philanthropy is an evolving phenomenon (Novelli *et al.*, 2016), yet despite having strong foundations within many forms of tourism (e.g. disaster, volunteer and eco-tourism), it remains significantly under-researched. As many tourists travel in search of the 'other' or the 'exotic', they commonly find themselves in developing countries, particularly those with an accessible indigenous population or fragile natural environments. Such destinations are often characterised by economic and social poverty, requiring the need, or perceived need, for TEFL tourists in order to attempt to help strengthen community development. It is poverty in this form that often makes a place 'foreign' and 'exotic' to materialistic eyes, with self-curated social media photos rendering such places 'romantic' and 'adventurous'. Many tourists romanticise the 'simplicity' of local lives, some develop a 'personal urge to help', others hardly seem to notice, while few feel a sense of unease when confronted with what Mostafanezhad (2013) calls 'geographies of compassion'.

Recent years have seen the West shift away from compassion for distant suffering towards a growing desire to act across those distances, to alleviate suffering directly and achieve physical and personal encounters with beneficiaries or 'distant others', while creating in-lifetime legacies (Silk, 2004). It is within such contexts that travel philanthropy has been provided with the opportunity to grow, although it can be argued that the very notion of philanthropic travel is multifaceted, theoretically nascent and ambiguous. Philanthropists have long aimed at achieving social transformation, and travel philanthropy in all its forms has evolved through the democratisation of charity, as a kind of 'doing good' through 'giving back' while travelling (Novelli *et al.*, 2016). TEFL tourism provides the perfect platform for travel-based philanthropic activities such as this to occur.

It has been argued that philanthropy is a niche in tourism but, to a similar extent, tourism is a niche in philanthropy (Bergman, 2013). Existing research largely focuses on corporate philanthropy or the sustainable practicalities of philanthropic travel but, to date, there appears to be little examination of philanthropic tourists themselves. With philanthropic or altruistic motivations forming an integral part of relevant academic debates, most notably in the volunteer tourism literature, this section briefly contributes to the currently under-developed body of literature addressing philanthropy in tourism.

Philanthropic behaviour broadly refers to all activity which promotes human well-being, for example blood donations, voluntary work or financial contributions (Reddy, 1980). In the context of tourism, Goodwin *et al.* (2009) posit this as the donating of money, in-kind resources or time occasioned or facilitated by travel. Similarly, such practices are also frequently referred to in the relevant literature as 'altruism'. Novelli *et al.* (2016) outline the three dominant types of philanthropy in today's postmodern society: *strategic corporate*, *social entrepreneurship* and *social justice* philanthropy. As previously stated, the corporate market monopolises much of the academic literature addressing travel philanthropy and, although there are some organisations which are involved with TEFL projects through corporate donations or staff volunteering schemes such as PepsiCo for example (PepsiCo, 2016), these are neither widely known nor are they the focus of this book. Instead, it is proposed that understanding strategic philanthropy undertaken by corporations may aid in the comprehension of individuals with possible philanthropic intentions, including TEFL teachers.

Also termed 'philanthrocapitalism' (Edwards, 2009), strategic corporate philanthropy is frequently utilised as a means of enhancing competitive identity and social engagement (Hero, 2001; McAlister, and Ferrell, 2002). Although organisations may have benevolent intentions, the potential for enhanced commercial prospects is frequently at the heart of their motivations, such as promotion of their brand image or increased potential to work with sponsors, as opposed to the cause they are claiming to support (Porter and Kramer, 2002). This same notion could be applied to TEFL tourists who exhibit motives such as career development or who undertake philanthropic endeavours as part of a course requirement, where helping others is secondary to the personal benefits they gain. Social entrepreneurship is the attempt to draw upon business techniques and private sector approaches to find solutions to social, cultural or environmental problems, and while this approach may differ from traditional businesses and have a greater focus on improving society, it is still liable to these aspects of self-enhancement, particularly where there is a lack of accountability (Wagner, 2002).

Social justice, or social change philanthropy, focuses on building a broad, progressive movement that aims to address the root causes of social and economic inequalities (Goldberg, 2002). It could be argued that the attempts of many developing nations at improving English-speaking capabilities and the ensuing recruitment of TEFL teachers could fall within this category, although it appears, in many instances, that commercial prospects are the dominant motivation behind the government's endeavours in this regard.

Goodwin *et al.* (2009) suggest the use of 'travel philanthropy' as an umbrella term for the three practices of individual giving (either directly or via an intermediary) and corporate and/or personal fundraising and volunteering. As previously discussed, volunteer tourism has typically been associated with selfless activities or behaviours concerned with the well-being of others. Scholars have, however, now begun to question its role (Wearing, 2001, Callanan and Thomas, 2005; Coghlan and Fennell, 2009), highlighting that while volunteer tourism will almost always consist of elements of helping others, it is not necessarily indicative of the volunteer tourist's primary motivations (Lyons and Wearing, 2012). Wearing and McGehee (2013) advance this debate, arguing that there may not be such a thing as an unselfish volunteer and that many volunteers will, in fact, disguise their search for self-discovery behind their altruistic facade.

Stebbins (2000) further suggests that altruism and self-serving needs should be placed side by side, indicating that one does not come at the expense of the other. Other researchers propose that volunteers can adopt any position on the continuum between pure altruism and pure egotism and can possess multiple motivations simultaneously (Hustinx, 2001; Tomazos and Butler, 2010). Lyons (2003), however, argues that taking a general approach to the extent to which altruism plays the role of a key motivator for volunteer tourism is a fallacy, since volunteers are not a homogenous entity. This thus supports Stainton's (2016) contention that the current research approach to volunteer tourism is too broad to facilitate true understanding of the specific types of volunteering available, for example TEFL teaching.

## Pro-poor TEFL Tourism

The prevalence of pro-poor tourism (often shortened to PPT) has increased substantially in recent years as a result of a growth in general societal consciousness and awareness of sustainability issues. This is true in many walks of life, not only in the tourism industry. Environmental, social and economic considerations have moved from the periphery to the forefront in decision-making for many consumers in recent years, and organisations have responded through the introduction of corporate social responsibility plans, sustainability policies or by making amendments to existing operations. In other cases, individuals or organisations have exploited this consumer consciousness, capitalising on the goodwill of tourists and commodifying philanthropy into pre-organised or packaged experiences. This can be seen within TEFL tourism.

As demonstrated through the studies examining motivations to take part in volunteer tourism, there is now a far greater push by tourists to 'do something good' or 'give something back', and PPT provides the perfect platform for doing so. The World Tourism Organisation defined sustainable tourism as early as 1988 as leading to the management of all resources in such a way that economic, social and aesthetic needs can be fulfilled while maintaining cultural integrity, essential ecological processes, biological diversity and life support systems. To date, much of this debate has centred around environmental sustainability or the enhancement of community involvement in tourism. While many initiatives may incorporate pro-poor elements, this dated approach to 'sustainable tourism' fails to take into account the poverty element. In 1998, the UK Department for International Development (DFID) commissioned DeLoitte and Touche, along with Dilys Roe of the International Institute for Environment and Development (IIED) and Caroline Ashley of the Overseas Development Institute (ODI), to report on the potential for outbound UK tourism to contribute to poverty reduction, where it was found that, while tourism already helped poor countries in many instances, there was most certainly scope for this opportunity to be capitalised upon. This was then subsequently addressed in 1999 at the meeting of the UN Commission on Sustainable Development, where governments were urged to maximise the potential of tourism for eradicating poverty by developing appropriate strategies in cooperation with all major groups, indigenous and local communities. Thus, the foundations for PPT were provided by placing poor people and poverty at the centre of the sustainability debate.

PPT is essentially tourism that produces net benefits for those living on the poverty line. PPT intends to increase the benefits for the poor from tourism and ensure that tourism growth and development contribute to the cause of poverty reduction. PPT is not a specific product or sector of tourism, but an approach that is prevalent in a number of tourism niches, including TEFL tourism. PPT strategies aim to provide opportunities for people from disadvantaged backgrounds. This could be for economic gain, other livelihood benefits or increased participation in decision making. Poverty reduction is not usually at the heart of a tourism agenda, yet tourism is an important economic player in many developing countries and is already affecting the livelihoods of millions of poor people, positively and negatively. PPT strategies are considered to include aspects such as, but not limited to, infrastructure gains, capacity-building and training, and empowerment (Goodwin, 2009).

While the main focus of PPT frequently tends to be on direct or indirect employment of local people, there are a range of potential benefits (and possible negative impacts) of this type of tourism that are yet to be fully examined and exploited. TEFL tourism is one such example. Here, the aim is not to provide the local community with financial rewards as a result of tourism through means of employment, but instead to provide opportunity for community development through increasing the English language capabilities in the community. This, in turn, provides the indigenous population with enhanced career prospects and the potential to earn more money than they otherwise would. This additional revenue in the local area can then help boost development levels through taxation and responsible government spending. While this may sound somewhat utopian, it is important to note that PPT, along with the wider tourism industry, is not without its critics. These issues are discussed further in Chapter Seven.

One of the major problems when considering PPT initiatives is the ambiguous notion of what constitutes poverty. Poverty is a multidimensional phenomenon that consists of a 'lack' of certain things. People are generally viewed as being in poverty when aspects of their life sacrifice their physical and mental well-being and human health, although there are no hard and fast criteria from which to assess whether one qualifies as being in poverty or otherwise. This is an ongoing debate in a general sense and is not limited in any capacity to tourism studies. Poverty is frequently defined in either relative or absolute terms. *Absolute poverty* measures poverty by the amount of money needed in order for basic needs to be met such as food, clothing and shelter. In contrast, absolute poverty is not concerned with broader quality of life issues or with the overall level of inequality in society. Indeed, this concept fails to recognise that individual people have particular social and cultural needs. Criticisms subsequently led to the development of the concept of *relative poverty*. Relative poverty defines poverty in relation to the economic status of other members of the society, meaning that people are poor if they fall below prevailing standards of living in a given societal context.

This causes an inevitable element of confusion when determining whether a tourism activity is classed as pro-poor or not, given the subjective notion of the term poverty itself. Social scientists are generally critical of the absolute and relative definitions of poverty, which are based predominantly on economic data. Instead, sociologists tend to study the reasons for poverty, examining aspects such as the roles of power, culture and social structure among other aspects. Many studies in this area are based around the premise that individuals are frequently influenced by the physical and cultural context in which they live, and this approach gives importance to gender and household structure (UNESCO, 2018).

While the international poverty line is useful for measuring progress according to specific dimensions of poverty, it is based on a reductionist conceptualisation of poverty (Overseas Development Institute and Associates for International Resources and Development, 1999). Nowadays this method of assessment is utilised less rigorously and instead it is widely felt that one cannot consider only the economic part of poverty, but that one must also take into account social, political and cultural aspects. Poverty is considered now by many to undermine human rights in an economic (the right to work and have an adequate income), social (access to healthcare and education), political (freedom of thought, expression and association) and cultural (the right to maintain one's cultural identity and be involved in a community's cultural life) sense (DFID, 2001). Although certain aspects of poverty may long continue in some temporal and geopolitical contexts, generally it is more useful to view poverty as a complex phenomenon that often includes some or all of the above dimensions.

It is important to note that the characteristics defining poverty are also frequently invoked as its causes. For example, a person's lack of education may be deemed the cause of poverty as well as the result. The causes of poverty can set in motion an unavoidable interlocking cycle, whereby people are unable to escape poverty due to the cause of poverty. Take English capabilities, for example. A person may not have a lot of money because they do not speak the language required for the jobs in the local area, but because they cannot secure a job, they cannot afford to take English lessons to improve their skills. This lack of money may make the sale of

**Box 3.8.** Do something good: volunteer teaching

This is an excerpt from the website GapYear.com (2018) on the benefits of TEFL tourism.

> Teaching English as a foreign language (TEFL) is a popular gap year option and volunteering as a teacher is a great way to gain classroom experience first-hand.
>
> If you're thinking about teaching as a career, TEFL is the perfect starting point; you can make a career in TEFL as itself! Also, it's one of the best ways to further fund your gap year. Travel, teach, travel, teach ... easy!
>
> There are all sorts of ways to volunteer as a teacher and there are huge benefits to the students. Often, learning English is one of the best routes out of poverty for kids in many parts of the world and even learning a few words can help them later on in life. So, if you think you've got what it takes to TEFL then jump in at the deep end – it's not something you'll forget!

material assets necessary in order for people to be able to afford basics such as food and shelter, therefore increasing a person's vulnerability to changing economic circumstances while also leading to a decline in their socioeconomic status. Therefore, poverty should be viewed as a process as well as a stand-alone phenomenon.

Interventions from others can help break this vicious cycle, providing those in absolute or relative poverty the opportunity to seek a better life. TEFL tourism is one such example, where donations and free or cheap labour provided by tourists can help to develop the language capabilities of a group or individual with minimal cost to them. However, this is based on the assumption that philanthropy is at the heart of the TEFL tourism industry. Unfortunately, this is not the case in all circumstances, with capitalist profit-orientated organisations and hedonistic or selfishly motivated TEFL tourists prioritising their own lives and experiences above their students. While there are evident links between TEFL tourism and PPT, coupled with ample potential for this to be the case, the dynamism of the industry makes this an impossible area to succinctly measure and assess. Chapter Seven delves more into such sustainability concerns.

**Box 3.9.** English speaking staff needed in Bangalore

The following is an extract taken from BBC (2017).

> In 2003 BT was the latest in a series of companies to relocate their call centres to India, where costs can be 30% lower than in the UK.
>
> The Indian call centre industry employs more than 100,000 workers, who, if serving a foreign market, are often trained in the culture and customs of the country they service.
>
> Workers serving UK customers are often given accent training, and taught about pubs, football and running story lines in popular soap operas, to be able to hold conversations with British customers.
>
> BT's Indian call centres will deal with tasks such as ringing UK customers to remind them to pay their bills.
>
> Whilst this caused considerable controversy in the UK, with many British workers being laid off, it provided English-speaking Indians with a vital employment opportunity and aided in boosting the economy of the local area.

## Educational TEFL Tourism

Education is fundamentally at the core of the TEFL experience. This is most notably the education provided by the teacher to the students, but also includes the education that the TEFL tourist receives in terms of skill development, learning about the local culture and general knowledge through travel. Further to this, many TEFL packages as sold by tour operators such as i-to-i and Real Gap include as part of their product a TEFL qualification, where the tourist is provided with the opportunity to learn both theoretical and practical aspects of teaching English as a foreign language. It can therefore be seen that TEFL tourism demonstrates several close ties with educational tourism.

The progressive development of the tourism sector and the introduction of new alternative tourism forms, such as TEFL tourism, coupled with changes in education, have seen the convergence of the tourism and education industries (Ritchie *et al.*, 2003). Despite the significant size and scope of both, there has been little academic regard for the educational tourism phenomenon and meagre discussion on its conceptual foundations and definition (Ritchie *et al.*, 2003; van't Klooster *et al.*, 2008; Falk *et al.*, 2012; Stone and Petrick, 2013;). Surprisingly, Ritchie *et al.* are the only scholars who appear to have attempted to define the educational tourist, drawing upon definitions of tourism and the parameters of educational tourism. Accordingly, an educational tourist is 'a person who is away from their home town or country overnight, where education and learning are either the main reason for their trip or where education and learning are secondary reasons, but are perceived as an important way of using leisure time' (Ritchie *et al.*, 2003, p. 18).

This definition is nonetheless limited in its approach since it does not account for the type of education, reasons for the education undertaken or who the education is for. The TEFL teaching experience has two distinct educational spheres: the education provided by the tourist and the education the tourist receives. To date, the limited research addressing educational tourism focuses on the tourist's cognitive development, as opposed to the people that the tourist may educate, in this case their TEFL students, which demonstrates a considerable gap not only within the literature on TEFL tourism, but also the broader educational tourism field.

Drawing on the general literature addressing the volunteer tourism industry as a means of beginning to understand the TEFL teaching phenomenon, it is apparent that education is a common participant motivation (Wearing, 2001; Brown, 2005; Coghlan, 2006; Benson and Seibert, 2009; Gecko *et al.*, 2009; Leonard and Onyx, 2009; Lo and Lee, 2011). References generally point to the volunteer's cognitive development and greater awareness of 'self' (Wearing, 2002; Lepp, 2008; Matthews, 2008; Guttentag, 2009; Sin, 2009; Wickens, 2011; Wearing and McGehee, 2013). Pertaining to the notion of learning through experience and self-development, there is evidence that volunteer tourism can help participants to exhibit philanthropic behaviour (Lo and Lee, 2011; Coren and Gray, 2012) and foster self-reflection and developments in personality traits and behaviours (Gray and Campbell, 2009; Broad and Jenkins, 2008; Soderman and Snead, 2008; Sin, 2009; Wickens, 2011; Alexander, 2012; Benson and Wearing, 2012). As with TEFL tourism, many volunteer tourism projects provide participants with the opportunity to have a 'grass roots' experience (McIntosh and Zahra, 2005; Palacios, 2010; Lo and Lee, 2011; Wickens, 2011; Coren and Gray, 2012; Grabowski, 2013), enabling greater educational attainment with regard to the host country and community (Raymond and Hall, 2008).

Self-development is not restricted to the experience itself and has been known to permeate beyond the trip, often transferring into the tourists' everyday lives (Brown, 2005; Lepp, 2008; Grabowski, 2013; Wearing and McGehee, 2013). Broad and Jenkins (2008), for example, identified that volunteer tourists in Kenya developed a new perspective on life once they returned home, coupled with an intrinsic need for meaning and purpose in their lives. It has also been known for those involved with international travel to develop skills such as problem-solving, time management and communication (Scarini and Pearce, 2012). Although the 'curriculum' of such informal learning may not be well organised, there is plenty of information to process and travellers are likely to acquire new skills and perspectives as a result (Pearce and Foster, 2007).

Despite the frequent references to the educational benefits of volunteer tourism, there is little known about how or specifically what the tourists learn. Within the TEFL experience, there are two clear distinctions between the types of TEFL teacher learning that takes place: formal (TEFL courses, teacher training) and informal (learning through travel or cultural immersion). Accordingly, Ritchie *et al.* (2003) suggest that the educational tourism industry should be segmented. The first proposed segment is university, college and school tourism, in which the tourist experience is secondary to formal learning and can be described as *education first*. The second is edu-tourism, defined as general travel for education and known as *tourist first* (Ritchie *et al.*, 2003). Although such categorisation does not appear to have taken place within the broad body of literature addressing the educational aspects of volunteer tourism, some programmes do form part of a course of study abroad (Lyons and Wearing, 2008), and this is a useful categorisation for TEFL teachers as many undertake courses to gain recognised qualifications as part of their TEFL experience.

Such segmentation may facilitate differentiation between tourists, but it is difficult to measure whether educational benefits are the result of travel, interaction with other cultures, classroom study or a combination of these. Abrams (1979) states that study abroad is better defined as 'learning through experience abroad'. There is an increasing body of research providing evidence that most learning takes place outside of the formalities of the education system (Falk *et al.*, 2007; Falk and Dierking, 2010), although to date there is little reference to how, what or why such learning takes place (Falk *et al.*, 2012). Pearce and Foster (2007) describe travelling as its own kind of educational institution, with the experiences and knowledge gained representing a parallel to formal education. Travel, it is argued, offers one of the few opportunities outside of formal education where non-vocational learning about other times, places and people takes place (Werry, 2008).

In order to demonstrate the relationship between travel and learning, some scholars have applied fundamental theories used in learning and education analysis to tourism research (Boydell, 1976; Broomhall *et al.*, 2010; Coghlan and Gooch, 2011; Falk *et al.*, 2012; Stone and Petrick, 2013). While it is beyond the scope of this book to undertake an in-depth theoretical analysis of the way in which learning occurs within the TEFL experience, it is clear that learning is a core component. From a theoretical perspective, the first noteworthy connection is with existential learning. This is in essence a form of education that takes place as a result of learners gathering knowledge for themselves, usually as a result of personal experiences, and can be defined as 'meaningful discovery' (Boydell, 1976, p. 19). Stone and Petrick (2013) contend that this factor can be used as a model for explaining how people learn by travelling. This can be particularly useful in the case of TEFL teachers who will inevitably learn from their experiences both inside and outside of the classroom. Kolb's (1984) model essentially outlines the way in which a person will have a concrete experience, reflect on that experience, conceptualise or conclude what has happened and then experiment or try out what they have learned before having a new concrete experience. This is a continuous cycle and is a useful way to attempt to understand learning as a result of travel (Fig. 3.4).

Building on the pedagogy of existential learning and, in particular, the reflective aspect, Coghlan and Gooch (2011) propose that volunteer tourism organisations should utilise Mezirow's (1991) theory of transformative learning to inform their operational plans. This theory describes a shift in one's assumptions and world beliefs through a series of steps and highlights that the individual experiences a deep, structural shift in the basic premises of thought, feelings and actions as a result of the learning experience (O'Sullivan, 2002). It is possible to apply such transformative learning to the practice of TEFL teaching. Scholars have identified educational tourism as being a form of lifelong learning, fostering the continuous development and improvement of the knowledge and skills needed for employment and personal fulfilment throughout life (Broomhall *et al.*, 2010; Falk *et al.*, 2012). This theoretical association is particularly relevant to the examination of TEFL tourism in that the experience consists of formal and informal learning, whether this be in a classroom environment, 'on the job' learning, or education acquired through living and travelling in another country enabling the teacher to gain transferrable skills and experience to aid their future development on both a personal and professional level.

## Adventure TEFL Tourism

As tourists continue to seek out new and alternative travel experiences, adventure tourism continues to grow in popularity and has become a major niche within the special interest tourism sector. Adventure tourism is a term without a clear definition. Indeed, different people have different perceptions of what might constitute 'adventure'. For one person, adventure may be comprised of simple activities such as camping or walking, while another person might consider this as passive tourism. For others the notion of

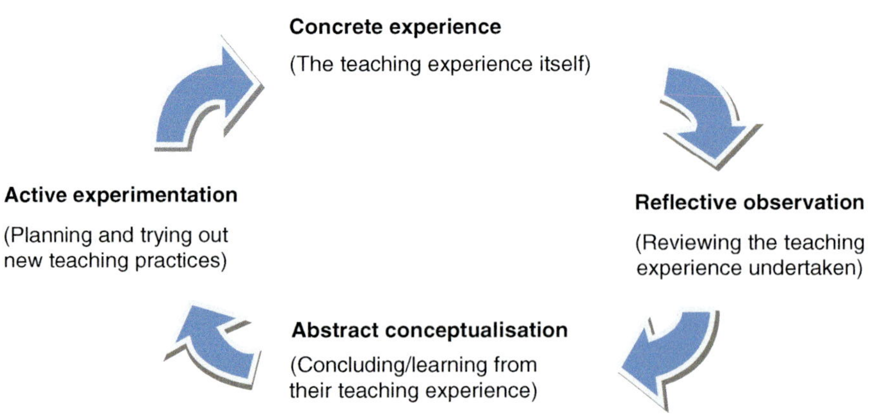


**Fig. 3.4.** Existential learning in TEFL teaching based on Kolb's (1984) experiential learning model

adventure tourism is associated with dangerous and physically or emotionally demanding and challenging activities, such as rock climbing or white-water rafting. According to the Adventure Travel Trade Association (2018), adventure tourism is 'any leisure pursuit which includes a physical activity, a cultural exchange, or activities involving nature'. A tourist does not necessarily have to go bungee jumping or scuba diving with sharks to be an adventure tourist; they could simply be connecting with a new culture through their classroom experiences as part of a TEFL tourism project. The stimulation and intensity associated with adventure tourism contributes to removing the experience from the routine of everyday life. Exotic surroundings, new activities and experiences all contribute to a sense of escapism and it can be argued that 'adventure' is a chance to set aside the everyday concerns of life.

Early adventure research was undertaken within the leisure science, adventure education and adventure recreation fields (Ewert and Hollenhorst, 1989; Hall and Weiler, 1992; Ewert and Shultis, 1997). Recently, research contributions have focused on the adventure tourism market, introducing the notion into the realms of special interest and niche tourism (Bentley and Page, 2001; Weber, 2001; Hudson, 2003; Cloutier, 2003; Ewert and Jamieson, 2003; Swarbrooke *et al.*, 2003; Wilks and Page, 2003; Page *et al.*, 2005; Williams and Soutar, 2005; Buckley, 2007). Adventure tourists are commonly young, educated, middle class, active thrill seekers who spend significant amounts of money in the pursuit of adventure (Tsui, 2000; Swarbrooke *et al.*, 2003). They can be demanding and discerning tourists and research has found that they traditionally travel to remote and extreme environments of the world to satisfy their needs for emotional highs, risk, challenge, excitement and novelty (Crompton, 1979; Bello and Etzel, 1985; Christiansen, 1990; Zuckerman, 1994).

There is an abundance of literature addressing participant motivations in adventure tourism and recreation. Buckley (2012) summarises these into three dominant categories of: internal performance of activity; internal/external place in nature; and external social position. The first category covers aspects such as thrill, fear, control, skill usage, achieving challenges, personal fitness and risk. The second comprises aspects such as appreciation of nature and beauty, artistic activity and spiritual experiences, and the third covers enjoyment of activities with others, enhancement of self-perception, notions of escapism and competition. These motivations coincide with research (Stainton, 2017a) examining the motivations of TEFL tourists in Thailand, where tourists exhibited the desire to utilise and develop specialist skills, undertake the challenge of TEFL, work with others (both colleagues and students), improve the perception that others have of them (particularly in terms of career and skill development) and to do something different from their everyday lives. Aspects such as fear and control are also likely to be encountered by TEFL tourists due to working in new, unknown and challenging environments, although this has not been explicitly addressed in research to date.

People are motivated to undertake adventure tourism activities for many different reasons. Some tourists enjoy the anticipation of an unknown or uncertain outcome by, for example, undertaking something new and unfamiliar or due to a perceived danger in the activity. Adventure tourism activities are generally associated with a degree of challenge alongside a perceived reward on completion of the activity. Intrinsic rewards associated with adventure tourism could be things such as reaching the summit of a mountain or helping TEFL students to achieve their desired grade in class.

**Box 3.10.** Elephant Camp and Hill Tribe Teaching TEFL Adventure

TEFL Adventure is a subsidiary of the company Enjoy TEFL, offering packages combining tourism activities alongside TEFL tourism. Here is an example of a product that they are promoting.

> The Elephant Camp and Hill Tribe Teaching is a project where tourists get to spend time with, as well as teaching English to, the local kindergarten and to the Mahouts, who look after the elephants. This project is excellent if tourists wish to combine working with elephants and teaching English. The size of the groups taking this project are small, which provides an intimate experience of the project.
>
> Price £732 includes Free TEFL and Mindfulness Courses.
>
> (Enjoy TEFL 2018)

## Gap Year TEFL Tourism

For many tourists, their TEFL experience forms part of a wider touristic opportunity, such as a gap year. Indeed, most gap year websites emphasise both voluntary and paid placements, such as TEFL, as a key option during this 'time out' period and academics have noted that this is a growing trend among gap-year travellers (Lyons and Wearing, 2008; Soderman and Snead, 2008). The past few decades have seen long-term independent travel grow exponentially, with the boom in the gap-year industry being most evident in the UK, where taking a gap year was first popularised. This tourism sector has steadily grown to become a significant form of outbound tourism supported by an industry of commercial and NGO providers (Simpson, 2004). Gap years have also become popular among other tourism-generating regions such as North America, Australia and New Zealand as well as many parts of Europe. Once a marginal and unusual activity undertaken by those described as 'hippies' and adventure-seekers, gap-year tourism has now become a widely accepted rite of passage for young people.

The terms 'gap year' and 'year out' are problematic because neither represents a tightly-defined phenomenon. Associating such years with ideals of freedom, personal development and fulfilment, many see a gap year period as an educational experience, or as a period of fun and independence before taking on the roles of responsible adulthood. Gap years tend to consist of a nominal period of time during which a person delays the next stage of formal education or employment in order to travel (Millington, 2005), although this interlude may be experienced at any point across the lifespan. Gap years are most commonly associated with early adulthood due to the popularity of young people taking a year off to travel after completing secondary school or tertiary studies.

Although there is a distinct dearth of academic research on this subject, existing literature covers issues such as youth travel and backpacker tourism. It is important to note, however, that neither academic nor industry-based literature

**Box 3.11.** Reasons to teach English as a foreign language during your gap year

The top reasons to undertake TEFL during a gap year according to gapyear.com (2018) are as follows.

**To make money**

If you are going to teach English as a foreign language in a developing country, you will be paid very handsomely in comparison to the national average wage of that particular country. English teachers are highly sought after and this is reflected in the pay. Your living costs will be minimal, meaning you can easily save money to further fund your travels and supplement your budget.

**To boost your career**

If you would like to pursue a career in teaching then TEFL can serve as a brilliant introduction. Many schools don't require a formal TEFL qualification, meaning you have nothing to lose if you just want to give it a crack. You'll experience what it is like to stand in a classroom in front of a load of kids and really get a feel for whether you enjoy it or not. If you do, having experience of TEFL on your CV will look fantastic and give you a leg up over the competition back home.

**To help others**

English is a route out of poverty for many people all over the world, particularly children. Even if you just teach them the basics, they will have something to go on and the drive to improve further. If people can speak English – aka the world's language – they will find it much easier to find jobs as tour guides and such like.

**To experience a country**

You will get to experience a country and all its customs fully if you work there for a little while. It's all well and good trotting around the tourist sites taking pictures, but nothing can absorb you into local communities in the way that working in them can. You'll see a whole different side than most people see to a country, giving you ultimately a more rewarding experience.

provides an adequate basis for understanding all of the different groups of tourists taking gap years or the diverse range of activities that they choose to undertake. Building on the discussion about macro-niches, gap tourism can be considered a macro, with a number of micro-niche tourism forms, such as TEFL tourism, beneath it. In order to fully comprehend gap-year travel, it is therefore imperative not to look at the bigger picture, but instead to consider the smaller, sub-industries and their particulars.

According to the Department for Education and Skills (DfES) (2004), gap-year trips can be distinguished by choices in broad types of activity. Structured activities refer to those set up, managed and facilitated by a host organisation. Forms of structure can vary, but most involve components of training and instruction before the tourist is placed on an organised scheme. This is in contrast to a wide variety of possible unstructured activities which the tourist undertakes on an individual basis. The DfES groups activities into six generalisations: work (paid and voluntary); learning; travel (organised and independent); and leisure (with individuals typically undertaking a mix of these activities during a gap year). Gap-year tourism describes a wide array of activities, some of which have been criticised as hedonistic (Simpson, 2004; O'Reilly, 2006).

Other forms of gap-year tourism, including long-term independent travel, have been equally criticised as a form of escapism and dropping out (Rojek, 1983). One major tour operator has evocatively branded a gap year as the 'Big Year Off'. Despite such perceptions and associated criticisms, there is a growing view that undertaking a gap year for travel purposes is not wasted time, and marketing in this area typically promotes the concept as more than simply an extended holiday. In Australia, for example, where school education finishes at Year 12, some promoters have gone as far as calling a gap year 'Year 13' (Youth Central, 2018). Also in support of this notion, one Australian university has offered a programme that enables prospective students to gain academic credit for learning demonstrated through the use of a reflective journal during their gap year (University of Canberra, 2009).

Simpson (2004) suggests that gap-year programmes encourage people to develop their global citizenry by 'broadening their horizons' through travel. Jones (2005) further argues that gap-year tourism can be transformative and that participants can benefit from the wider learning associated with cross-cultural experience. Desforges (1998) suggests that in some cases, the cultural capital young people gain from these experiences is converted into economic capital

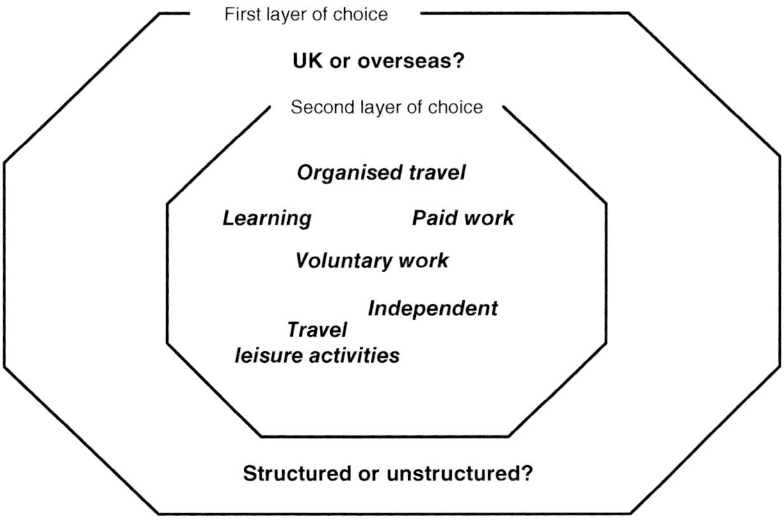


**Fig. 3.5.** Model of choice in gap year activities (DfES, 2004)

when they return home, and Heath (2007) dubs the gap year as the 'economy of experience', whereby the experience gained works as an important means of gaining the edge over others in a competitive marketplace.

As demonstrated throughout this chapter, TEFL tourism has close ties with a number of macro-niche tourism forms, which have been subjected to research and scrutiny in recent years. From these neighbouring tourism niches, the TEFL industry can learn many lessons. Alongside this is the concept of the micro-niche tourism form. Due to the size and scope of micro-niches, such as TEFL tourism, there is far less known about the day-to-day running of the industry or its management. Other relevant examples include linguistic tourism and service learning. Linguistic tourism, typically associated with the learning of a language which is foreign to the tourist, can be deemed closely associated with TEFL tourism, albeit the roles are reversed and it is the tourist who teaches the language as opposed to learning it. Although for some tourists learning the local language may make up part of their internal motivation, it is unlikely to be the primary reason for undertaking a TEFL tourist placement. Likewise, the tourist may constitute their TEFL experience as a form of service learning, but this is probably not the primary function.

Regardless of the extent to which TEFL tourism may or may not relate to these concepts, it is important to recognise the sparsity of literature in both of these areas. While they have had some attention within the academic community, this has been extremely limited. Thus, it is more sensible instead to take on board the associations between TEFL tourism and its macro-niche counterparts, given the credibility of developed research in these areas. Further reference will be made to the various tourism forms outlined in this chapter throughout this book, as a means of managing the TEFL tourism industry and ensuring mitigations of unsustainable practices previously evidenced in alternative tourism sectors.

# 4

# The Commodification of TEFL in the Context of Tourism

The world we inhabit today is very different from the world in which we lived only a few short years ago. Societal demographics and psychographics have changed and continue to change in response to consumer demands, along with the political and economic climate. Across the globe, political and economic systems are becoming saturated by thirsty, private, for-profit organisations. This impacts the TEFL industry in two ways: the increased demand for English-speaking capabilities in emerging economies and the commodification of the TEFL product.

The first point is the driver of the boom in the current TEFL industry. The social system within which we now live, coupled with the development of new technologies and transport systems, has facilitated the growth of business in parts of the world that were previously underdeveloped. The economies of nations such as China and India are now growing at exponential rates as a result of capitalist society and its forthcomings. Increased globalisation and international trade requires the use of a common medium within which all parties involved can communicate, most notably English. With this comes the demand for English language improvements on a significant scale in these localities and thus this is a strong driver for the TEFL tourism industry.

Despite capitalism being at the heart of the boom of the TEFL tourism industry, however, this is not the focus of this chapter. Instead it examines the progressive changes seen within the TEFL industry and the way in which commercial, public and NGO bodies operate within it. While the ideals of TEFL may lie in the likes of philanthropy and community development, this is now being replaced by profit-orientated and hedonistic motivations.

## TEFL in a Capitalist World

Capitalism is an economic and political system in which a country's trade and industry are controlled by private owners for profit, rather than by the state. Private entities, which tend to exercise control through companies, are provided with the opportunity to own the factors of production – i.e entrepreneurship, capital goods, natural resources and labour. Capitalism requires a free market economy to succeed, in which the market sets the prices of the components of supply and distributes them according to the laws of supply and demand.

As demonstrated in Chapter Two, the demand for English language communication, and by default therefore for English language teachers, has increased exponentially in recent years across the globe. Meanwhile, supply in many instances has not been able to meet such demands. In economic theory, the law of supply and demand is considered one of the fundamental principles governing an economic market. It is described as the state where, as supply increases, the price will tend to drop or vice versa, and as demand increases the price will tend to increase or vice versa. While prices to take English classes may in some circumstances have increased according to the law of supply and demand, what we have seen in the TEFL industry is a step away from traditional means towards an alternative method of teaching English as a foreign language: TEFL tourism.

The very prospect of capitalism, which has now spread across the majority of the globe, has enabled individuals to exploit the TEFL marketplace, developing business plans according to profit incentives and commercial opportunities as opposed to the focus on community development that governments or public bodies often have. This has enabled the deconstruction of the traditional supply and demand equation, where the aim is to achieve equilibrium between price and quantity, resulting in a multi-tongued industry that has become increasingly complex. The root of the problem lies in the context of economic and political development from a geographical point of view. Countries have progressively fought for years for better living conditions, democratic freedom, equal rights, better economic opportunities, etc., and ingrained in these concepts is the ability to communicate in English.

While this book largely addresses TEFL within the context of tourism, there is in fact far more depth to the history of, and reasoning for, English language development in particular parts of the world. It is not the purpose of this book to provide detailed histories of Marxism or world systems analysis, but it is important to understand the links between these theories and TEFL in order to adequately conceptualise the TEFL tourism industry within today's capitalist marketplace.

### TEFL and globalisation

Globalisation has very much been at the centre of international development for the past few decades. It is the sense of connectivity in economic and cultural life across the world that has been growing for centuries. According to Robertson (1992, p. 8), globalisation is 'the compression of the world and the intensification of consciousness of the world as a whole'. Essentially it makes the world a single place, which gives rise to unilateral questions about the running of societies, economies and political systems (Fig. 4.1). This confrontation of world views means that globalisation involves comparative interaction of different forms of life (Robertson, 1992).

Contemporary society was largely shaped as a result of international communications, transportation and conflict dramatically intensifying relationships across societal and geographic boundaries. During this time, the main reference points of fully globalised order took shape: nation state, individual self, world system, societies and 'one humanity'. These elements of the global situation were viewed under a new light, where national societies and individuals began to interpret their very existence as part of a larger whole as opposed to the singular view that they once had. To some extent, a common framework has guided this – for example, the development of regulatory guidance or support such as the World Trade organisation (WTO) or the United Nations (UN).

It is this globalisation, accelerated by new technology, which has had the greatest impact on the English language. English is no longer a small island language as it once was, it is now a world language spoken in all corners of the globe either as a mother tongue or for purposes of business and commerce. Most notably, English is the dominant language of the sciences, computing and throughout academia among other industries. People do not necessarily need to have travelled to have been exposed to the English language. Tremendous changes in international travel and tourism and in the growth of the internet have made the world a metaphorically smaller place, one that is increasingly accessible once people can communicate in English.

The growth and development of English as a global language is one of the most significant

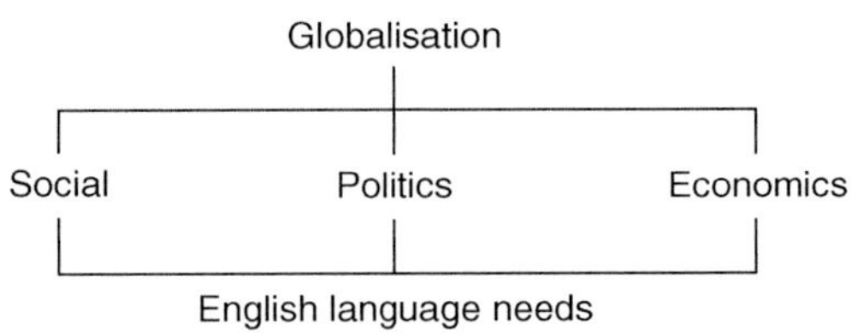


**Fig. 4.1.** The link between globalisation and English language needs

**Box 4.1.** Advert for online English tuition

Below is an advert from the UK sales website Gumtree.

**ENGLISH LESSONS ON SKYPE – ONE TO ONE – PRONUNCIATION – EXAM PRACTICE**

**Basingstoke, Hampshire**

Do you ever feel that you lack confidence, or are nervous about joining in with English conversations, writing emails or letters? If so, then I can help you.

My name is Rob, I'm British and I have almost twenty years teaching experience. I have a BA (Hons) degree in Education and Training and I also have a Trinity Cert TESOL qualification. I am offering Skype/Online tuition in:

English conversation – this will enable you to speak with confidence and listen with clear understanding. The focus of the lesson will be to improve the English skills you need in order to communicate with confidence.

I design lessons to suit your personal needs and your level of English.

I teach General English from Beginner to Advanced levels, and if required I can prepare you for PET, FCE, IELTS exams and also ESOL Entry 1 to Level 2.

If you don't want a qualification then I will teach you to speak with accuracy and fluency. As a native speaker tutor I can also teach you the meaning of idioms and phrases of English language that are not always taught in the books. Along with helping you achieve fluency in English, I will also teach you about English social life and English customs, so you will have no problem going out and making friends with English people.

My teaching can also help you with job interviews, work projects and college assignments. The first on line lesson is FREE and during this lesson we can also discuss your specific requirements.

The cost for each 60 minute lesson is £15.

I look forward to hearing from you and helping you reach your English language goals, so please contact me by email for further information, and you may also book your first lesson.

My Skype name is rbcroberts2

Add me to your contacts and let's talk English!

(Gumtree, 2018)

events of the late 20th and early 21st centuries. A single language has become effectively universal, allowing for it to be used as a global lingua franca for communication between speakers of many different languages. In addition to the language spoken in the UK, USA, Australia, etc., global development has seen the introduction of new varieties of English, often referred to as 'New Englishes' which have emerged from contact with local languages in the same way as Spanish, French and Italian evolved from Latin. In effect, 'world English' might have been no more than a celebration of diversity, like world Music, rather than the global lingua franca which it has also become.

The evidence points to a growing tolerance of multiple standards in English and growing flexibility and fluidity in the use of English by global citizens. While this may be beneficial in some contexts it is problematic for the non-native TEFL teacher who may teach versions of the English language not compatible with local linguistics. Native-speaker varieties of English may not be easily intelligible and strict adherence to a particular standard of English may not be the most important means of achieving mutual comprehension. Meanwhile, native-speaker features can cause problems for learners and those which are not essential for international intelligibility can be safely disregarded. So it can be seen that it is not necessary to sound like a native speaker in order to be understood around the world and speakers of global English do not have to give up their existing identities and approaches.

Governments across the world have recently embarked on ambitious educational reforms which integrate English more deeply into the school curriculum. For some, English will cease to be a foreign language and instead be repositioned as a 'basic skill'. A 'wave of English' has been building up across the globe for some years. Although English currently appears to be in an unassailable position in the modern world, its future as a global language is not guaranteed. The British Council has predicted that the number of people learning English is likely to continue to increase in the immediate future, but will soon begin to decline as a result of pressures to preserve national identities, resistance to colonial-based imposition of an unwanted language and changes in international powers (the USA has traditionally been the dominant power and therefore has brought with it the English language). Increasingly, the long-term future of English as a global language probably lies in the hands of Asia, and especially India and China, where English is seen as a common second language by huge populations with diverse dialects.

## TEFL and world trade

Being able to communicate in English is an increasingly important business skill as world markets become more globalised and interconnected than ever before. As new economies develop rapidly, the value of being able to speak Mandarin, Arabic and Hindi is increasing, but for now English remains the leading international language. The dominance of English is evident throughout many aspects of life including scholarship, the internet and technology, global media and advertising, international politics and diplomacy, and particularly, global business.

Growth in international commerce has brought together people from all cultures and countries in order to exchange goods and services and to facilitate trade. The economic incentive to learn English is compelling as an increasing number of international companies such as Nokia, Renault, Heineken and Samsung have recognised the long-term advantages to productivity and growth that adopting English as a common company language can have. Adopting the use of English enables businesses to be competitive by being able to communicate with a diverse range of customers, suppliers and other partners. In the current global climate, companies that fail to devise a language strategy are limiting their growth opportunities to the markets where their language is spoken, putting themselves at a disadvantage to competitors that have adopted English-only policies. In addition, language differences can cause a bottleneck when geographically dispersed employees have to work together to meet corporate goals. Organisations which provide a common medium by which staff can communicate are likely to benefit from smoother and more translucent processes than may otherwise be the case.

One of the core aspects of international business is the ability to communicate effectively. When stakeholders are unable to communicate in the same language, meaning and understanding are easily lost, whether this be face-to-face, on the telephone or via email exchanges. To further add to this perplexity, the role played by

**Box 4.2.** English language teaching in China – as effective as it seems?

China appears to be making significant efforts to improve the English language capabilities of its nationals through the integration of compulsory English classes in public schooling.

However, while it may appear on the outside that English language development is occurring, people working on the inside have reported otherwise. Behind the claim that 300 million people are either learning or have learned English in China is a reality that classes are frequently extremely poor and the teachers themselves have poor English capabilities. Rote memorisation is common practice and students are not taught to speak for themselves outside of the context of their textbooks.

So in reality, while it may appear that China is improving the national rate of English speaking at a rapid pace, lessons are in fact frequently insufficiently taught with learning taking place at a minimum.

(Based on *Economist*, 2011)

cross-cultural integration can be notoriously problematic, with subjective interpretations and colloquialisms making translation difficult.

On the other hand, one-language policies can have repercussions that *decrease* efficiency. Learning to communicate in another language can come as a shock for some and be a difficult barrier to overcome. Attending meetings conducted in a new language, speaking with those who are fluent and being required to produce written communications can have detrimental effects on employees. As a result, an erosion of personal self-confidence can have negative impacts on productivity and thus implications for the overall success of the business. While being required to communicate in a language in which one is not confident may have detrimental effects on the employee, it can also cause significant frustrations for the stakeholders of organisations who may feel that they cannot communicate effectively with the individual.

While working to improve English language capabilities is one thing, assessing these skills is a different matter. Gauging fluency is an extremely difficult task, largely owing to the subjectivity of the term. There have been various attempts to overcome this, with formalised processes put in place to 'measure' English language capabilities, as much as this is possible. One such example is the Common European Framework of Reference for Languages (CEFR). This is an international standard for describing language ability, depicting an individual's level on a six-point scale, from A1 for beginners, up to C2 for those who are 'fluent' (Fig. 4.2). The premise of this framework is to make it easy for anyone involved in language teaching and testing, such as teachers or learners, as well as for employers and educational institutions, to easily compare qualifications and examinations across different countries.

While formalised structures such as this may well facilitate an easier understanding of English language capabilities, this is not practical in all contexts. Those working in a call centre in Mumbai, for example, are more likely to have developed their English language capabilities in an informal context or in a formal context, but without formal exams. This makes the ability to measure English language skills once again a challenging task.

## The business of English language

While there are most certainly issues around English language speaking in relation to globalisation and international trade, there is no doubt that it is an imperative and essential part of the

**Box 4.3.** How to improve call centre English

Call centre agents working for companies in which English is the primary language used for business need to have excellent English-language skills. These skills are imperative to communicate effectively with co-workers, management and clients. Agents with poor or limited skills in areas such as active listening, pronunciation or business vocabulary can make mistakes that result in lost customers, billing errors or financial losses. Below are some methods to improve the English skills of call centre agents and reduce these issues:

1. **Work with a corporate trainer who is certified to teach English to speakers of other languages and skilled at teaching business English.** The trainer may help you build a curriculum customised to your agents' needs and organise appropriate training.

2. **Give staff access to the training materials and tools outside of training sessions so they can study on their own.** This can help to improve their abilities and confidence.

3. **Make changes around the call centre that focus on improving English skills outside of scheduled or self-development training.** For example, team leaders may deliver speeches or conduct meetings in English or staff may be paired together so that the stronger English speaker can help the weaker English speaker.

4. **Offer recognition to agents who successfully improve their communication and English skills.** You may present a plaque or acknowledge an employee's accomplishments with a bulletin board announcement or award a financial bonus.

(Chron, 2018)

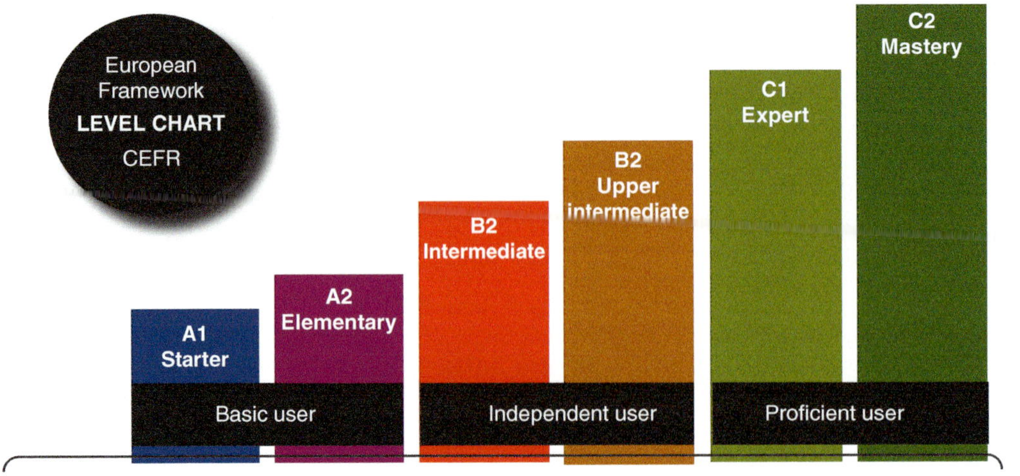


**Fig. 4.2.** The CEFR for languages six-point scale (Council of Europe, 2018)

success of many businesses across the globe. This chapter thus far has largely assessed how English language has transformed the world of business, however it is equally important to note, in the context of TEFL tourism, that business has also transformed the world of English.

Although in essence the product of English language is intangible, aligned with the ideals of enhanced social development and business prospects, there is a strong body of evidence demonstrating that, in many instances, it has been commodified into a product for sale in an ever-competitive capitalist market. In many contexts, language is now treated as an economic resource, cultivated for material profit, or acquired as a skill to be offered on the market (Park and Wee, 2012).

Commodification is the expression used to describe how a specific object or process is rendered available for conventional exchange in the market. While the concept lays its foundations in Marx's idea that capitalism was founded on the notion of turning work into a commodity, the word 'commodification' itself is recent, dating from the mid-1970s. Capitalism is centrally about producing and distributing commodities and has become increasingly normalised in today's society. History has demonstrated that budding entrepreneurs have pushed the boundaries of what can be turned into a commodity and the TEFL tourism industry is no exception.

Capitalism as a profit-making activity has traditionally existed in the shape of merchants and money-lenders acting as intermediaries between consumers and producers engaging in simple commodity production. What is specific about the capitalist mode of production is that most of the inputs and outputs of production are supplied through the market (i.e. they are commodities) and essentially all production is in this mode. For example, many TEFL tourism elements are operated and controlled by the TEFL agency acting as a tour operator or intermediary. Consequently, the whole organisation of the production process is redefined according to economic rationality as defined by the capitalist marketplace, expressed in price relationships between inputs and outputs (such as wages, production costs, sales, profits) as opposed to the larger rational context faced by society overall.

In essence, the whole TEFL process is organised and reshaped in order to conform to commercial logic. It has become a business. The growth of the capitalist mode of production has expanded to such an extent that it has now begun to infiltrate markets previously naïve to commercial exploits. As a result, the social organisation of TEFL has been refashioned and reshaped in a commercial (profit- and market-oriented) way; the 'old ways of producing' (for the primary purposes of social and economic development and to improve livelihoods) have now been displaced by this new industrialism.

Commodifying TEFL in the context of tourism activity has enabled it to become the product. This is represented through the process of combining various material inputs (transport, accommodation, etc.) and immaterial inputs (plans, know-how) in order to make something for consumption (the output/TEFL product). Many organisations and individuals are now capitalising on the demand for English language by setting up new businesses, franchises or subsidiaries where the teaching of English is sold on the respective market.

Although the commodification of English language will inevitably have positive outcomes for some (most notably entrepreneurs and those at the top of host organisations), socialists have railed against the market economy as inherently exploitative. Marx argued that under capitalism all value is produced by labour, which is forced by the class of capitalists to work longer than is necessary to provide for their subsistence and, over time, for their 'customary' standard of life as the economy grows and wages rise. The capitalist exploits the labouring class by appropriating the surplus value produced by labour, which is the value or output in excess of the value of their wages. This surplus value is then distributed as profits, interest and rent, and the surplus can be reinvested in expanding production which leads to economic growth.

In effect, Marx argues that capitalism creates an inevitable level of exploitation within society, where those at the bottom of the chain benefit least from the commercial exploits in which they are playing an often important or key role. In a traditional sense, this is demonstrated in TEFL through the commodification of the TEFL product, where for-profit organisations or individuals have capitalised on the demand for English language capabilities. Based on this theory, those employed to teach in language institutions or schools on low wages, unpaid voluntary teachers

**Box 4.4.** TEFL Cambridge Groupon deal

The following is an advertisement taken from Groupon (2017).

> With more than 35 years of experience in the industry and as an institutional member of the IATEFL, TEFL Cambridge offers ACCREDITAT-accredited training courses that are developed by professors and tutors. Lessons are designed with the most recent methodology and are carefully crafted to be fun and easy to understand. After finishing a unit, participants are required to answer a multiple choice test to assess their understanding of the material and as a way to prepare them for the examination. TEFL Cambridge believes that by increasing students' confidence, they will turn out to be skilled educators with internationally-valued qualifications.
>
> **The Deal**
>
> - £49 for a level 5 140-hour TEFL online course (88% off)
>
> **The course includes:**
>
> - The fundamentals of English teaching: structure of language I grammar foundations I teaching grammar practically
> - Teaching knowledge test (TKT) preparatory module
> - Language module: functions, grammar, lexis, and pronunciation
> - Language skills module: listening, speaking, reading, and writing
> - TKT Certificate 1 module: learning and teaching languages preparation for University of Cambridge TKT certificate 1
> - TKT Certificate 2 module: teaching materials and lesson planning preparation for University of Cambridge TKT certificate 2
> - Classroom management module
> - Teaching kids and teenagers: the basics for teaching younger students I managing the challenges I win them over with stories I young learners lesson planning
> - Teaching advanced students: English in business I ESP (English for specific purposes) I teaching individuals I test preparation
> - Teaching with imagination: handling large class numbers I adapting to less ideal environments I prioritising learning I understanding cultural differences
>
> (Groupon, 2017)

and most notably those who indeed pay to give their labour, are potentially being exploited in that they may often be working long and arduous hours while reaping little financial reward, whereas those at the top could be making a substantial sum of money as a result.

Unfortunately, this notion of exploitation is fuelled by the way in which TEFL tourism is promoted and indeed perceived by many people. Sentiments of goodwill and altruism are at the heart of many a participant's motivations and thus financial rewards are not high on the agenda. Little do they know, however, that those at the top may be reaping rewards on an unprecedented scale. This is addressed further in Chapter Six.

Referring again to Marx's theory of capitalist exploitation, it is important to note that he identifies exploitation in a wider context than that of employee wages. This, however, is difficult to assess within the TEFL tourism sector as it is an emerging industry with a lack of available specific data regarding financial input and output, although the assumption can be made that the growth of the industry is likely to have the said economic effects. What is clear, however, is that there are two spheres within which TEFL tourism promotes economic growth, first through the development of TEFL enterprises and second through the growth of alternative businesses as a result of improvements in English language capabilities. Therefore, economic improvements which result indirectly from the growth in TEFL tourism can potentially contribute to the exploitation of society on a broad level as described by Marx. It is suggested that only by transcending it and moving to socialism can such exploitation be overcome.

## The TEFL Business Model

While the evidence of commodification within TEFL is indisputable, the scene is rather hazy, with an abundance of different business models and approaches utilised throughout the industry. Some organisations may be genuinely charitable whereas others may only pose to be so inclined. Some make no claims to be charity ventures, but instead focus on the rhetoric of tourism, gap-year activities and fun. Some companies may be small independents and others may form a part of a multinational chain. Some are government-funded, others are funded by charities or sources which are not obvious, particularly to the untrained eye. In fact, it is relatively difficult to make much sense of the current marketplace owing to a lack of transparency.

In terms of ownership, TEFL businesses vary considerably. Some are owned by a single person or a small group of people and some are owned by large numbers of shareholders. TEFL organisations can also be owned by charitable foundations or trusts and some are owned by the state (either in the country where the TEFL is taking place or by an aid-giving nation). A business's ownership structure determines the legal responsibilities that it may have, including the necessary paperwork to set up the business, any relevant taxes, how profits are managed and distributed, and the business owners' personal responsibilities and liabilities if the business makes a loss or goes bankrupt.

At a broad level, it is possible to distinguish between TEFL organisations that are owned and run by private owners, those that are owned and run by the state and those that are run by

**Box 4.5.** The growth of Chinese sellers on Amazon

Improvements in English language capabilities have opened up new doors for people who were previously unable to communicate in English. One of the biggest growth areas is seen in retail, where many vendors have now broken into previously inaccessible marketplaces. E-commerce is now global and the geographic boundaries separating merchants from consumers have been overcome through the use of the internet and the ability to communicate effectively.

An article in *Forbes* magazine (*Forbes*, 2017) claims that since 2015 the number of Chinese merchants found on Amazon has increased tenfold. They are now provided with the opportunity to ship their goods to the likes of the USA and Europe and many have seen their business potential increase exponentially.

Naturally, increase in money owned by individuals then ripples into the economy, with those people increasing their daily expenditure in most instances.

voluntary organisations. Legal forms and ownership structures of businesses are different from country to country, an added level of complexity given the geographical nature of TEFL tourism. This book addresses TEFL tourism from the view of the UK, where the majority of businesses are sole traders, limited companies or business partnerships.

A sole trader is a person who is running a business as an individual. This is most common for those undertaking their own tuition, although some sole TEFL traders may employ staff to expand their business or to help with administrative duties. Sole traders can keep all the business's profits after paying tax, but they are personally responsible for any losses, paying the bills incurred by the business (e.g. rent or transportation costs), and keeping a record of all sales income and expenditures.

Most TEFL entrepreneurs are inclined to set up a limited company to run their business. Setting up a limited company provides legal rights and obligations in itself. These are separate from the rights and obligations of its owners as individual persons. For example, a limited TEFL company can own property to use as a language school, its finances are separate from the finances of its owners and any profit made after taxes belongs to the company as opposed to an individual.

Limited companies typically have 'members' or shareholders. They also frequently have 'directors', who may or may not own a share of the company. The role of shareholders and directors is to hold responsibility for the company's financial liabilities, such as losses or debts. There are two types of limited company: private limited companies and public limited companies. Public limited companies (also known as PLCs) will trade business on the stock market, enabling anybody to purchase shares in the company. This is different from private limited companies, where people can only buy shares with the approval of the business owners (e.g. if they are invited to invest in the company).

Some TEFL organisations operate on the basis of a business partnership. This is an arrangement where at least two individuals will share ownership of a business. Partnerships tend to come in two forms: general partnerships and limited partnerships. In a general partnership all partners are personally responsible for the business, which means that they are personally liable for any losses or debts incurred. In contrast, a limited partnership offers a level of personal protection as partners are not personally liable for any losses or debts incurred by the business. Profits from a partnership are shared between the partners and each partner then pays taxes on their share. There are a lot of specific details that can vary according to the structure of the business partnership which are important to identify when setting one up.

Legal and ownership structures, business size and industry sector are not completely separate from each other. Most sole traders, for instance, are small businesses which run independent tuition. This is because a single individual is unlikely to have the financial capacity to expand and run a large business, nor are they likely to have the desire to be personally liable if a business were to run into financial troubles. Within the tourism industry, there are large companies such as TUI and Thomas Cook which expand their businesses through the use of subsidiaries focusing on TEFL.

TEFL tourism organisations can largely be divided into two categories: those that are teaching-focused and those that are tourism-focused. On another level, agencies can be privately owned, NGOs, government bodies or charities (Fig. 4.3). The private sector is typically dominated by tourism-focused enterprises, many of which are large conglomerates, whereas teaching-focused businesses can be associated with both for-profit and not-for-profit organisations.

An NGO is any non-profit, voluntary citizens' group which is organised on a local, national or international level. NGOs are task-oriented and are normally driven by people with a common interest, such as improving English language capabilities in a given locality. NGOs perform a variety of service and humanitarian functions, bring citizen concerns to governments, advocate and monitor policies and encourage political participation. NGOs involved with TEFL typically provide analysis and expertise in the areas of community development and education and serve as early warning mechanisms as they help monitor and implement international agreements.

A society is composed of three sectors: government, the private sector and civil society, excluding businesses. NGOs are important components of social movements within a civil society.

**Box 4.6.** TUI shares

TUI AG (formerly TUI Travel plc, formerly First Choice Holidays) is the world's number one tourism business and owns the subsidiary i-to-i, a company which hosts a number of TEFL package-style trips. As of February 2018 the group umbrella consisted of strong tour operators, 1,800 travel agencies and leading online portals, six airlines, more than 130 aircraft, over 300 hotels with 210,000 beds, 10 cruise liners and countless incoming agencies in all major holiday destinations around the globe. This integrated offering enables TUI to provide 30 million customers with holiday experiences in 180 regions. A key feature of the corporate culture is global responsibility for economic, environmental and social sustainability, which is reflected in more than 20 years of commitment to sustainable tourism.

**Today's buy and sell prices**

Offer – 1,578.50p
Bid – 1,577.50p
Spread – 0.06%

**Today's trading**

Range – 1,569.05p–1,583.00p
Volume – 0.29m
Previous close – 1,579.00p

**Yearly summary**

52wk range – 1,068.00p–1,687.50p

**Fundamentals and health**

Market capitalisation – £9,274.84m
Shares in issue – 587.39m
PE ratio – 8.87
EPS – 178.00p
EPS growth – 178.13%
ROCE – 8.02%
Quick ratio – 0.77
Current ratio – 0.79
Total dividends per share – 63.00p
Dividend yield – 4.12%
Dividend cover – 2.74

(This is Money, 2017)

**Box 4.7.** Love TEFL business structure

Love TEFL is part of the TUI subsidiary i-to-i. It was founded in 1994 when it claimed to have pioneered the first TEFL trip 'changing the industry forever'. In 2001 it introduced its online TEFL course and by 2015 it had awarded 175,000 TEFL qualifications.

Although owned by TUI, there is no evidence of this on the website or promotional material, perhaps because Love TEFL wishes to disassociate itself from the mass holiday market image. It does, however, promote its relationship with gap-year company STA travel. STA is also a subsidiary of a large holiday provider, Diethelm Keller Holding Ltd.

Love TEFL was given a considerable head start over smaller competing TEFL tourism organisations given that its parent company has a near-monopoly of the travel and tourism industry in the UK. With the know-how and expertise of experts in the field, Love TEFL has grown to be one of the largest TEFL tourism providers in the UK.

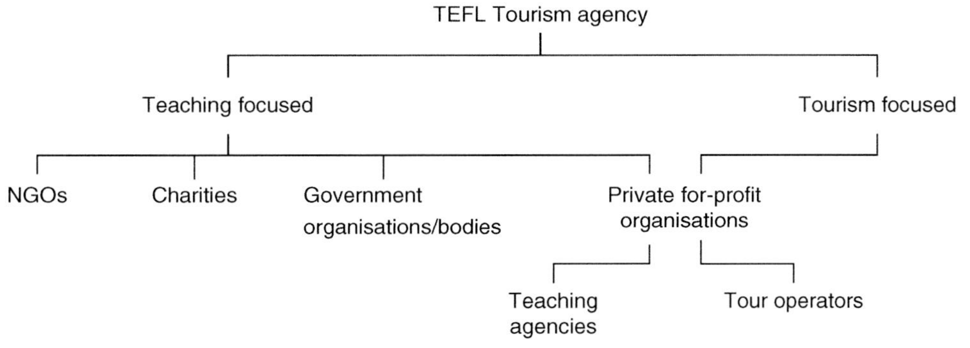


**Fig. 4.3.** Types of TEFL tourism agency

The issue of independence is an important one when determining the credibility of an NGO. It can be a challenge for NGOs to remain independent of government influence and it is common for individual governments to influence the NGO community in a particular field by establishing NGOs that promote their policies. This has been recognised by the use of the acronym GONGO which refers to a government-organised NGO. Also, in more authoritarian societies, NGOs may find it very difficult to act independently and may not receive acknowledgement from other political actors even when they are acting independently. However, development and humanitarian relief NGOs often need substantial resources in order to run their operational programmes, so many readily accept official funds which may play a role in the influence of policy and practice by the government.

The structures of NGOs within TEFL vary considerably. With the development of technology and improvements in communications, more locally-based groups, known as grass-roots or community-based organisations, have become active at national and global levels. This typically occurs as a result of the formation of coalitions with other NGOs for particular goals. NGOs can be divided into two groups: operational and advocacy NGOs. Operational NGOs attempt to mobilise resources, frequently in the form of financial donations, materials or volunteer labour, in order to sustain their projects and programmes. This process requires often complex organisation of finance obtained from grants or contracts, relevant liaison with governments, foundations or companies and time and expertise spent on planning, preparing applications, budgeting, accounting and reporting. Operational NGOs are frequently associated with major fundraising events. Advocacy NGOs, on the other hand, undertake much the same functions, but with a different balance between them. Fundraising takes place on a smaller scale and persuading people to donate their time is still necessary. Contrary to operational NGOs, external donors may not impose onerous administrative burdens, but supporters still have to be supplied with information on an efficient, regular basis. Major events will aim to attract favourable publicity rather than raise funds.

There are many charitable organisations which involve TEFL programmes, although the focus tends to be around broader developmental issues, of which improving language capabilities and education are important elements. A charity is essentially an organisation which is set up to help society. In order for an organisation to be classified as a charity in the UK, its aims have to fall into categories defined by the law as 'charitable', such as the prevention or relief of poverty (which TEFL can assist), or the advancement of the arts, culture, heritage or science. A charity cannot make profits for distribution, and all money goes back towards furthering the charity's causes. A charity also cannot have owners or shareholders who benefit financially from it. In the UK it is necessary for charities to clearly state their objectives in order to be registered with the Charity Commission. They must also explain how they meet these objectives in their annual reports which are available to the public.

**Box 4.8.** The School Fund

The School Fund is a not-for-profit NGO that fights to tackle the global education crisis. It believes that greater participation in secondary school supports economic growth, gender equity and better health. According to the organisation, every additional year of education increases a student's lifetime earnings by approximately 10%. It has a particular focus on gender, where it claims that the benefits of educating girls are particularly high: women who have gone to school have fewer unwanted pregnancies, healthier children and face lower infant mortality rates.

The School Fund works with a network of organisations, ranging from high-performing private schools to community-based organisations that support students attending local public schools. Field partners identify high potential students who are economically disadvantaged and these students are offered scholarship opportunities. The students are then promoted on the website so that donors can crowdfund their scholarships. The School Fund also tracks students' progress through school.

(The School Fund, 2018)

One of the key requisites for gaining charitable status is that the business has to be established for what is known as 'public benefit'. The intention of public benefit is to determine which organisations are eligible to become registered charities. To be eligible, a company has to do positive things and any negative side-effects or consequences should be offset by the positive outputs of the business. They must be of benefit to the public, or a sufficient section of the public, whether this be localised to a particular geographic area or a population with specific characteristics, such as people who cannot speak English. The Charity Commission uses its guidance and case law to decide whether an organisation passes the 'public benefit test'.

Charities are not owned by any particular people. A charity is controlled and its assets are held in trust by a board of trustees. These trustees are primarily responsible for ensuring that the charity is run effectively so that it is able to deliver its charitable purposes for the public benefit as set out in its aims and objectives. There are four main types of charity: charitable companies limited by guarantee; trusts; charitable unincorporated associations; and charitable incorporated organisations (CIOs).

A charitable company limited by guarantee is the most common type of charity that you will come across. Companies which are formed on this basis are limited liability companies. The activities undertaken by the charity are governed by the articles of association and, if based in the UK, companies are registered at Companies House. This type of charity has its own legal personality and is subsequently able to agree contracts with other organisations. It can also hold property in its own name. This structure is well recognised and having limited liability provides a level of protection for the charity members and trustees.

Unlike a charitable company limited by guarantee, a trust can be a simple and cost effective method of setting up a business. This charity type is governed by a trust deed, the contents of which are more flexible as there is no statutory framework to work within. As a trust is not a legal entity in its own right, it is unable to enter into agreements in its own name. Instead, trustees are required to enter into such agreements in a personal capacity, which of course brings with it a range of personal liability risks.

An unincorporated association is another simple and cheap structure for a charitable business. A charity of this type is governed by a constitution which is not prescribed. Similar to a trust, this structure has no limitation of liability and is not a legal entity in its own right. Therefore the members of the management or executive committee would enter into any agreements on behalf of the charity and would be personally liable.

A CIO is the newest of the legal forms for a charity. It is incorporated and regulated entirely by the Charity Commission. A CIO is governed by a constitution prescribed by the Charity Commission. It is not as flexible as a trust or unincorporated association. A CIO benefits from the fact that its members have limited liability and it has its own legal personality, providing the trustees with a level of protection.

**Box 4.9.** Oxfam Education Programme

Every day in developing countries, teachers provide quality education in difficult circumstances. Each day, however, more than 72 million children, many of which are girls, still miss out on an education.

Education, especially for girls, significantly improves children's chances of escaping poverty, but ironically poverty is the main reason that they miss school. Many parents cannot afford the fees, books and uniforms that are required and many local communities do not have the money to run a school.

Oxfam work to help people build and equip schools and train teachers to staff them and fight against attitudes and beliefs that prevent girls from completing their education. They also help communities lobby for government support to get more children into local schools, and to hold governments accountable for the delivery of quality education.

(Oxfam, 2018)

A government agency, often an appointed commission, is a permanent or semi-permanent organisation in the government that is responsible for the oversight and administration of specific functions. There are a variety of agency types which are normally executive in character. Government agencies are most often constituted in an advisory role and in the UK are either executive agencies answerable to government ministers or non-departmental public bodies answerable directly to one of the parliaments or devolved assemblies of the UK. They are frequently referred to as 'quangos' – quasi-autonomous non-governmental organisations. This concept is often applied in the UK and to a lesser degree in Australia, Canada, the USA and other English-speaking countries.

The government agency most relevant to TEFL in the UK is the DFID. This agency is responsible for leading the UK's work in the battle against extreme poverty. It tackles the global challenges of our time including destitution and disease, mass migration, insecurity and conflict. While its work is not necessarily focused on TEFL, English language capabilities are an integral part of international development, as demonstrated earlier in this chapter. As such, there is an inevitable crossover between the practice of TEFL and that of the DFID in some capacity.

The private sector is the segment of a national economy owned, controlled and managed by private individuals or organisations. The private sector has the ultimate goal of making money and developing business opportunities. A private sector organisation can be created by forming a new enterprise or by privatising an existing public sector organisation. A large private sector corporation may be privately or publicly traded. The private sector tends to employ staff through individual business owners, corporations or other non-government agencies covering a wide range of industries including education and tourism, both of which host the concept of TEFL tourism. Staff employed by said organisations are paid with part of the company's profits and as a result private sector workers may have more pay increases, more career choices, greater opportunities for promotions, less job security and less comprehensive benefit plans than public sector workers. While in many regards this is a positive, it is the private sector which fuelled the rise in capitalism, which as noted earlier, has its pitfalls as well as benefits.

Private organisations can be very healthy in an economy as the more they earn the greater the amount they pay in taxes. Taxes raised are then eventually spent back on the host community where money can be invested in areas of need such as healthcare or education. Such corporations have the capacity to grow at a fast rate and their profit maximisation motive means that they place a strong focus on improving productivity and thus contribute to the gross domestic product (GDP) of the country as much as possible, which in turn often leads to a higher standard of living.

As a private organisation works primarily for its own interest, it has the opportunity to isolate itself from all worldly matters (unlikely for a business wishing to expand) and do what is best for itself. The owners have the free choice of making investments and decisions which will generate

**Box 4.10.** English language teaching programme procurement advert

The UK's Department for International Development (DFID) is considering procurement for a new Education programme in Burma. If approved it is anticipated that the programme would be expected to mobilise in the next 12 months, and would be perhaps a 5 year contract.

The unique political and economic environment in Burma makes this a particularly exciting programme for both DFID and our supply market to work on. We would be seeking one or more implementing partner(s) to implement:

- The design and implementation of activities to strengthen the education system in Burma.
- The provision of technical assistance to education systems strengthening, including management and planning as well as curriculum development.
- Strengthening English language education, including through teacher education colleges and through supervision and support to teachers at school level.
- Support to non-state education services, including through ethnic education systems. Identification and support to implementing partners who can provide education services in areas of the country emerging from conflict.

(DFID, 2017)

**Box 4.11.** Building an ESL business

Amy Lightfoot built her TEFL business from scratch. Here are her suggestions for doing the same as published by the British Council (2018a).

**1.** Figure out what you're good at and what you want to do.
**2.** Get to know people. Speak to people at conferences, get some business cards made, and follow up leads, no matter how small.
**3.** Don't forget your own professional development. Investing in a new training course always seems to lead to something new.
**4.** Look for links with other fields and industries. The skills you develop when working in English language teaching relate to so much more than just private language schools or global publishers. Being freelance, despite the occasional frustration and unpaid holidays, really is about being free; you're free to make your own choices and steer your own path.

the highest revenue for their business. They would not be obliged to consider any external factor (such as welfare) while designing policies. The problem here lies, however, in the nature of TEFL tourism and questions can be asked regarding the moral and practical aspects of converging the notion of international development with goodwill at its heart with those who are profit incentivised. This is addressed further later in the chapter.

In the TEFL tourism marketplace, private organisations tend to fall into one of two categories: those that are teaching-focused and those that are tourism-focused. Teaching agencies may be a means to facilitate TEFL tourism, but they are not actively involved in any tourism-based elements of a teacher's trip. While a teaching agency will organise and manage the act of teaching, similar to employment agencies in other industries, a tourism-based agency offers a package-style experience typically inclusive of teaching practice along with other elements such as transport, accommodation, excursions, qualifications, etc.

Some teaching agencies exist to help ease the job search process, both for the school and for the teacher. They help to find a school or organisation that is best suited for the teacher's needs and preferences. This includes looking at aspects such as geographical location, school types or specialised subject areas. Teaching agencies will then typically continue to assist teachers during the transition and throughout their period of work for that employer. They assist with visas, CRB police checks, bank accounts, etc., provide teaching tips and liaise with the school should there be any

**Box 4.12.** SEE TEFL overview

SEE TEFL is a for-profit organisation based in Thailand. It offers TEFL certification equivalent to a Cambridge CELTA and Trinity Cert TESOL, training in teaching English to children, Thai language classes, Thai culture training and a job guarantee.

SEE TEFL has successfully trained over 1,400 TEFL teachers and its accreditation is based on organisations which have the actual authority to certify and licence organisations for international quality management standards (recognised by governments), and permission to teach and train in Thailand (from the Thai government).

The school is licensed by the Thai Ministry of Education to teach TEFL courses, as well as English and Thai language courses and it was the first TEFL training organisation in Asia to achieve ISO 9001:2008 quality management certification, the backbone for maintaining quality throughout the organisation and in all of the language and teacher training courses, as well as teacher placement. It is an Approved Centre of Training Qualifications UK (TQUK), which in turn is recognised by the UK government Office of Qualifications and Examinations Regulation (Ofqual).

The course fee is currently set at USD $1,395, but it is not clear how this money is subsequently distributed/spent by the company.

concerns about the employment. Sometimes agencies will also help to secure accommodation.

Essentially, this type of agency is the intermediary between the teacher and the educational institution. They coordinate employment, contracts and often salaries on behalf of the school. Such agencies often charge an administrative fee, either to the school or to the teacher, or both. On the one hand, this may cause concerns over the use of a 'middle man' in this way – for example, the agency may not have a good understanding of the school's requirements or may not match teachers with schools adequately. However, employment agencies of this type can be very helpful in the TEFL industry due to the geographic and sometimes isolated location of educational institutions; they can help overcome cultural barriers or communication difficulties and make arrangements (i.e. for accommodation) on behalf of the teacher that may otherwise be difficult or stressful.

Other teaching-based agencies tend to take the form of language institutions, also known as language centres or language schools. These generally facilitate private language tuition, either on a one-to-one basis or through group classes. They vary significantly and can be large institutions, multinational chains, franchises or small independent businesses. Profit margins for such businesses vary widely.

A tourism-based TEFL agency is very different from the teaching-based agencies discussed above. They tend to align themselves with the tour operator model, combining tour and travel products to make a 'package'. A tour operator is responsible for operating and providing for a vacation through contracting, booking and packaging together various components such as hotels, transportation, meals, guides, optional tours and sometimes flights. Many TEFL tour operators can be considered land-only operators as they focus on the aspects of the package once the tourist has arrived at the destination and therefore do not typically include international flight arrangements.

According to Sharpley (2006), a package holiday is simply defined as the pre-arranged combination of two or more components of a holiday, such as transport, accommodation and other services (such as excursions). Thus, although tour operators are mostly associated with the organisation of charter flights to summer sun or ski destinations, it is important to recognise that there is an enormous variety of types of package holiday, particularly given the growth in special interest tourism such as TEFL tourism.

The rapid growth in the online marketplace has further complicated the chain of distribution while limiting the power of intermediaries. Many online businesses (dot.coms), such as expedia.com or booking.com, sell a variety of travel and tourism products, allowing customers to create their own package holiday (known as 'dynamic packaging'), while principals themselves – for example, budget airlines – also offer links on their websites to other products, such as accommodation, car hire, insurance and entertainment. This has changed the face of the tour operator industry, demonstrating a

**Box 4.13.** British Study Centres franchising

British Study Centres (BSC) is an English language school based in the UK. It is part of the British Study Centres group of colleges.

BSC operates a scheme whereby high quality schools of English worldwide may apply to become a British Study Centres Worldwide (BSCW) franchise.

As a franchise, schools are entitled to brand themselves as BSCW and issue certification to students accordingly. Naturally, this is subject to the maintenance of required quality thresholds, however the scheme is designed to be flexible enough to allow for the particular needs and parameters of local markets worldwide.

English schools that seek BSC status do so for a variety of reasons and therefore in order to accommodate these differing needs BSC offers a 'menu' style franchising package that enables prospective franchisees to select the level of support that suits their particular needs.

For example, an existing school in a particular country might already have an excellent academic base with strong teaching systems and pedagogical methods in place. Such a centre might simply seek to acquire the marketing and branding benefits of association with a respected UK-based English language teaching institution. Other existing schools may seek to upgrade their academic structures by learning from a peer organisation in the UK. New schools on the other hand may seek franchise status for the provision of a complete 'set-up' service covering everything from launch marketing to the establishment of appropriate staffing and pedagogical structures.

BSC's 'menu'-based franchising is designed to enable duly approved franchise schools to buy into as much or as little of the available BSC franchise service as they wish (subject only to their school meeting the required quality thresholds for admission to BSC status).

(British Study Centres, 2018)

progressive move away from the well-known multinational chains such as TUI and Thomas Cook, towards a more independent and individualised marketplace.

In recent years consumers have changed, becoming increasingly demanding in their requirements and new and emerging tour operators have capitalised on this opportunity. Many TEFL tourism organisations satisfy the tourist's desire to 'do something different', while providing them with a level of comfort through the packaging of their products. These tend to be small, independent tour operators or subsidiaries of larger companies, such as Love TEFL who are owned by TUI (Box 4.7, p. 68).

## The TEFL Package

Traditionally the mass tourist industry has generally been perceived as the antithesis of philanthropic tourism, a notion many may see as an important part of TEFL. However, as identified throughout tourism realms, the boundaries between tourism forms, including the dichotomies of niche and mass, have become increasingly blurred. This is demonstrated through the study of TEFL organisations and the TEFL products available on the market where, despite TEFL tourism being a micro-niche, it associates itself comfortably with the mass tourism model through the packaging and commoditisation of the TEFL product in a form closely aligned with the traditional package holiday model (Stainton, 2018a).

Holiday packages have historically been an integral part of the mass tourism sector. Most commonly facilitated through the use of third party agents, often referred to as tour operators, a number of components are purchased and assembled to make it convenient for the customer. Elements most commonly include transportation and accommodation and there may be others such as insurance or excursions. Although the fundamentals of combining elements into a commodified product in this way have remained the same for the past 25 years, the nature of packages has evolved in situ with the dynamicity of the tourism industry, as Vainikka (2014) explains. Often considered a manifestation of mass tourism, package holidays have traditionally been associated with sea, sun and sand destinations and along with the homogenous and standardised nature of Fordist mass (Poon, 1993;

Fletcher *et al.*, 2013). Despite the progressive move away from this association, mass and alternative forms of tourism largely remain dialectically polarised and have been described as hierarchical (Vainikka, 2014): mass tourism is associated with negative connotations and alternative or sustainable tourism forms are viewed as 'good' or 'better'.

Weaver (2007) argues that alternative forms of tourism, such as TEFL tourism, have begun to converge with the mass market, but the process is asymmetrical and heavily skewed towards mass tourism. This is demonstrated through the significant increase in the number of commercial operators that have subsequently changed the face of niche tourism industries, such as volunteer tourism. Although there do not yet appear to be any academic studies focusing on the fiscal nature of the TEFL industry in Thailand, the commercial presence of the industry in the market is indisputable. A 2018 Google search conducted by the author for the term 'TEFL teaching abroad' revealed 385,000 results, demonstrating the significance of this sector. To examine all opportunities available to the consumer, however, is far beyond the reach of this book. Stainton (2018a) provides insight into the scale of such opportunities in the summary of the opportunities in Thailand as presented on the first page of results, providing an overview of the TEFL opportunities advertised and highlighting the commercialisation of the TEFL sector (Table 4.1).

It can be argued that the organisations listed in Table 4.1 act as postmodern tour operators that have moved beyond the traditional stereotype of the package holiday, as noted by Vainikka (2014). As demonstrated in the table, alongside traditional elements of accommodation and transport, many TEFL programmes offer support such as travel advice and orientation prior to departure, visa assistance and optional excursions (e.g. orphanage tours). While, in essence, an inclusive package such as this may replicate the essential components of a modern package holiday, the nature of TEFL teaching is very different from the typical sea, sun and sand holiday. Despite this fundamental difference, the websites of the organisations listed promote a strong theme of tourism, with images and rhetoric of exotic beaches, cultural tourism, camaraderie and parties. This factor highlights the need for the conceptual amalgamation of package tourism and TEFL and gives rise to the question as to why the two research areas have not yet been combined by scholars, despite the strong associations as presented to the consumer.

## Tourism Rhetoric

The success of TEFL tourism operations is largely dependent on the ability to market and promote the product on offer. Demonstrated in Chapter Three, the links between TEFL and tourism are indisputable. Tourism rhetoric is most obvious in the marketing material presented by TEFL agencies, where location appears to frequently take precedence over the activity.

In many cases, TEFL tourism marketing focuses on attracting people to the TEFL destination before it addresses the act of TEFL. For some locations, the destination and its attractions are so well known, the tourism marketeer simply needs to remind prospective consumers of the appeal. In others, the tourist attractions and appealing factors are highlighted through aspects such as the website or social media content. Typically, such content is visually depicted, whereas the TEFL activity tends most frequently to consist of written content, either alongside or without accompanying images to represent the TEFL experience.

Images evoke emotion and instantly convey more than written content alone. Vision is the dominant human sense, providing a powerful platform from which organisations can market their product. Sight takes up to 50% of the brain's resources, thus making people literally 'hardwired' to respond to images. While there is an argument to say that text also is visual, images do the best job of reaching into a person's brain and affecting the intuitive and emotional inhibitions. While many may believe that TEFL tourism should be predominantly focused on the act of teaching English, it is the destination in which TEFL takes place that actually persuades many consumers to sign up to a programme.

There are a number of approaches that organisations can take when assessing the best way to market their product. One such is through the use of 'activity marketing'. Some TEFL agencies market themselves through the

**Table 4.1.** TEFL opportunities in Thailand

| TEFL organisation | Placement details | Cost to TEFL teacher | Monthly salary in country | Person specification | Programme synopsis |
|---|---|---|---|---|---|
| i-to-i (owned by First Choice Holidays) | 5 months teaching<br>Orientation and full support throughout<br>120 hours online TEFL training and 2 days classroom training | £1095 | £315* + accommodation and weekday meals | Must hold a university degree<br>Must be a native English speaker<br>No experience required<br>Must be aged between 21–45<br>Must have a passport from the UK, Ireland, USA, Canada, Australia or New Zealand | '… discover the ins and outs of laid-back Thai life, by living as an English teacher in the local community. Explore lively cities, chill out on idyllic white sand beaches, visit serene temples and get to grips with teaching English in a friendly host school. You'll earn a generous monthly allowance and be able to live very comfortably, and you'll receive great support throughout.' |
| TEFL Heaven | 6 months + teaching<br>3–4 weeks face-to-face TEFL training<br>Guaranteed placement hosted via Media Kids | £1095 | £600* + accommodation during TEFL training | Must hold a university degree<br>Must be a native English speaker<br>No experience required<br>Must have a passport from the UK, Ireland, USA, Canada, Australia, New Zealand or South Africa | 'Join a face-to-face TEFL course with around 20 or so like-minded individuals, train in TEFL, with complimentary holiday resort accommodation in a paradise location in Thailand – and receive a guaranteed paid teaching job afterwards for 6 months or more! Get 3 times more than the average local salary and enjoy a country that is known internationally as "The Land of Smiles!"' |
| International TEFL Academy (promoted via Goabroad.com) | 6–12 months teaching<br>180-hour TEFL course + 20 hours practical training<br>Lifetime job placement assistance | £935–£1450* | £350–£500* | Programme open to worldwide participants<br>Must possess excellent native-level English<br>No experience necessary | 'Teaching English overseas not just an opportunity to work abroad, it is the chance to truly immerse yourself in a new culture, experience a different way of life, build friendships that will last a lifetime and discover your own potential … you'll have the opportunity to experience a land, its people, and culture firsthand, see the sights and sample the local cuisine, and make friends you'll have for a lifetime … International TEFL Academy can be the gateway to making your travel dreams a reality.' |

| | | | | | |
|---|---|---|---|---|---|
| Teach English ESL | 5–10 months + teaching<br>120-hour TEFL course<br>Lifetime TEFL support | £1660* | Salary depencent on qualifications | Must be able to read, write and speak English fluently (no need to be a native speaker)<br>No experience necessary<br>No qualifications necessary | 'If the limestone karst formations of Thailand's southern coast have you beaming with delight, our **Krabi TEFL Course** is just for you. A long-time favorite destination for beach goers and rock climbers alike, with stunning sunsets, island hopping and great nightlife, Krabi has all the charm of a beach city and none of the unwanted fluff. With our **three-week, internationally accredited TESOL/TEFL course in Thailand**, you can live your dream of teaching English abroad. And the best part of all? It comes with **ESL job placement** in Thailand!' |
| SEE TEFL | 4–5 months teaching<br>2 weeks teaching essentials training<br>SEE TEFL Certification awarded on completion of semester one + 2000-word essay | £900* | £500*+ accommodation during training | Must hold a university degree<br>Must be a native English speaker<br>Must be aged 21–50<br>Must have a passport from the UK, Ireland, USA, Canada, Australia or New Zealand<br>Must possess police clearance from their home country<br>No experience necessary | 'The SEE TEFL paid Internship is an ideal way for people who wish to experience living and working in Thailand as a paid English language teacher, but who are not ready to commit to a 4-week training course, or a teaching commitment longer than 5 months ... with a salary of not less than **25,000 Thai Baht (THB)** per month ... It is possible to save money from this salary during the internship period, and afford at the end of the teaching placement a few months travelling Thailand and Southeast Asia and/or sitting on a beach under a palm tree, or trekking in the jungles ... There will be other foreigners in the internship placement town, but these are likely to be other teachers, adventurous travellers or expats.' |
| Teach Away (2016) | 12 months teaching<br>Online TEFL qualification available separately | Not stated on website | £300–£600* | Must have a university degree<br>Must have a TEFL certificate<br>Usually will have previous ESL teaching experience | 'For those who are interested in teaching English in Thailand, Teach Away offers a variety of teaching jobs ranging from ESL instructors in private language institutes to English teachers at private international schools. These positions are ideal for teachers looking to advance their professional careers, or for social, active individuals looking to travel and explore new employment opportunities in a dream location.' |

*Continued*

**Table 4.1.** Continued.

| TEFL organisation | Placement details | Cost to TEFL teacher | Monthly salary in country | Person specification | Programme synopsis |
|---|---|---|---|---|---|
| Gap 360 | 6–12 months teaching<br>120-hour TEFL/TESOL course<br>1 week Thai language and culture orientation | £999 | Up to £830* | Must be aged 21–35<br>Must be a native English speaker<br>Must have a passport from the UK, Ireland, USA, Canada, Australia, New Zealand or South Africa<br>No experience necessary<br>No qualifications necessary | 'This programme is unique as it offers a whole week of exciting cultural immersion, where you can discover all about Thai culture, see the sights, learn a little of the Thai language and have fun trying Thai cooking or visiting temples! Get a paid teaching job in Thailand after training and earn up to USD $1200 per month in a teaching placement. Train directly in a Thai school environment, gain practical classroom experience and study at our beachfront training centre with views of the sea in beautiful, laid-back Hua Hin! Meet like-minded mates and bond with teaching buddies as you train. Make a real difference to the lives of Thai kids as a school teacher, plus volunteer for 2 days teaching at a kids camp during your TEFL training!' |

Based on i-to-i (2017), TEFL Heaven (2017), Goabroad (2016), Teach English ESL (2016), SEE TEFL (2018), Teach Away (2016), Gap 360 (2016)
*Based on May 2016 exchange rates.

nearby attractions that TEFL tourists can visit as part of their placement/TEFL experience. This may include, for example, local beaches, animals and cultural attractions. It will also often include events and festivals that occur in the local area. As identified by Stainton (2017d), many TEFL tourists are interested in nightlife activities and therefore this may be something that TEFL tourism organisations wish to promote.

Another popular marketing approach is through the use of corporate marketing. There are a range of TEFL events and conferences held in various localities across the globe. Many TEFL organisations will also approach schools, universities and colleges and attend community events to promote their products.

Basic marketing addresses four pillars for creating and selling a product or service: product, price, place and promotion. In TEFL tourism marketing, the four Ps can be applied in the following ways:

### Product

TEFL tourism marketing includes determining the unique selling benefit or benefits one location has over its competition. For people looking to combine business and pleasure, a destination might offer ease of travel to and from the area, sufficient accommodation options, interesting nightlife, and activities for partners and children. Prospective TEFL tourists will also want to know about the TEFL provision which is included, such as any training and qualifications, job offers and ongoing support.

### Price

Fundamentality, TEFL tourism has two spheres when it comes to price. The first is the cost of the TEFL programme. Participants will naturally be seeking the best value for money with the most reputable operator. Secondly, TEFL tourists will be interested to know what remuneration benefits are associated with their placement. Alongside these considerations there may be an assessment of value for money and cost of living in the particular location in which the TEFL tourism is due to occur.

### Place

The 'place' refers to where a business distributes its product or service, such as in a store, online, using catalogues or through wholesalers. In TEFL tourism this is more and more often being done through tour operators and online travel agents, handled largely over the internet. Aspects such as search engine optimisation and domain authority will therefore play an inevitable role in the 'place' in which the TEFL tourism product is being promoted and most organisations will recruit specialist staff in this area.

### Promotion

Tourism marketing uses a wide variety of communication strategies and techniques to promote areas and destinations. TEFL tourism organisations might purchase advertisements in trade magazines or on websites. They may also promote themselves through paid advertising on search engines. Forums are also a popular place for marketing, particularly 'electronic word of mouth' (EWOM).

As demonstrated throughout this chapter, the business of TEFL tourism is one which is growing rapidly and which requires due care and attention like any other business. Increasingly resembling the traditional tour operator model, there are most certainly practices, both good and bad, which can be implemented within a TEFL tourism business or learned from.

Increasingly, the nature of TEFL is becoming more and more removed from its philanthropic foundations, moving towards the commodified and commoditised macros of the niche tourism industry. With this comes notions of capitalism, subsequently questioning the very nature of TEFL tourism. Commonly a for-profit business, TEFL tourism is now a fast-developing industry, which requires adequate monitoring and management to ensure its sustainability. This is covered in subsequent chapters of this book.

# 5

# The Education Paradox

There is a distinct paradox between the concepts of education and tourism, raising the question 'should tourism be used as a method of providing education'? While intentions may be good, tourists are not necessarily the best fit for the education sector. They may not possess the required skillsets, motivations or intentions required for the development of education in the way that is intended. However, turning to tourists may in fact be the only solution for some destinations, but this does not mean that it is necessarily the best solution. This chapter addresses the issues associated with the education sector and its convergence with the tourism industry.

## Developed vs Developing-world Teaching

Education systems vary considerably across the world, the biggest differences being between the rich and the poor nations. In most countries, education is compulsory for six years, for children from age 6 to 11. Primary schools occupy most medium-sized villages and above, but it is common for older students to have to travel a considerable distance from their home to a middle or high school. According to a UNICEF report in 2012, all children must have access to primary education that is free, compulsory and of good quality, although, seven years later, progress on this front appears to be rather slow.

While TEFL makes up only a small proportion of the education provided in developing countries, it is nonetheless a core ingredient of many schooling systems, particularly in countries where there is a push to improve the overall English language capabilities of the nation. Educational programmes typically adopt traditional western models of education, with an emphasis on language alongside maths, science and social studies. Western staff may or may not be involved with such educational institutions and the standards of schooling between destinations and institutions can vary considerably.

The problem lies, however, with the adoption or 'copying' of western approaches, which may not necessarily be aligned with the values and cultural practices of the local community. Many educational programmes in developing countries, for instance, allocate scarce resources to topics like Greek mythology, prime numbers or tectonic plate movement – topics that may provide intellectual stimulation, but have little relevance in the lives of impoverished children. What such systems frequently fail to recognise is that high-performing students in less developed regions face a very different future from their counterparts in wealthier areas: they are more likely to work on family or neighbourhood farms or start their own small enterprises. As a result, schooling provides neither the financial literacy students need to manage what resources may be under their control in the future, nor the

**Box 5.1.** UN Millennium Development Goal 2: achieve universal primary education

The aim of this goal was to ensure that, by 2015, children everywhere, boys and girls alike, will be able to complete a full course of primary schooling. The UN states:

- Enrolment in primary education in developing regions reached 91 per cent in 2015, up from 83 per cent in 2000.
- In 2015, 57 million children of primary school age were out of school.
- Among youth aged 15 to 24, the literacy rate has improved globally from 83 per cent to 91 per cent between 1990 and 2015, and the gap between women and men has narrowed.
- In the developing regions, children in the poorest households are four times as likely to be out of school as those in the richest households.
- In countries affected by conflict, the proportion of out-of-school children increased from 30 per cent in 1999 to 36 per cent in 2012.

(United Nations, 2018)

**Box 5.2.** Educational outcomes in India

In India, national survey evidence states that only about one-third of children in grade 5 cannot perform long division, and one-third cannot perform two-digit subtraction. Nearly one-half of grade 5 students cannot read a grade 2 text and one in five cannot follow a grade 1 text. Sixty percent of Indian children enrolled in grade 8 cannot use a ruler to measure a pencil. Only 27 percent of Indian children who complete primary school can read a simple passage, perform division, tell time, and handle money, although students should master each of these skills by the end of the second year of school.

(UNESCO Centre for Global Development, 2013)

guidance needed to create opportunities for securing a livelihood or building wealth. What's more, this approach to schooling provides little assistance to promote the physical health needed for economic stability and quality of life in many parts of the world.

From the perspective of TEFL tourism, there is every possibility that the very presence of a western tourist in said educational establishments can exemplify this situation. A lack of understanding of local needs coupled with ethnocentric perceptions of educational provision is likely only to make the situation worse. UNESCO (2012) points out that what students in impoverished regions need is not more academic skills, but rather life skills that enable them to improve their financial prospects and well-being, including financial literacy and entrepreneurial skills; health maintenance and management skills; and administrative capabilities, such as teamwork, problem-solving, and project management. It is more than possible for such topics to be integrated into TEFL teaching, should the TEFL teacher be made aware of local priorities and values.

In most developing countries, few children complete their secondary school education, and many do not finish primary school. There are several reasons, including the difficulty of getting to school due to great distances or lack of transport options, and the cost of schooling. Even when there is no fee for tuition, there are often expenses – for example, lunch, uniforms and examination fees. In many cases, the quality of education is not viewed as being to a high enough standard, requiring parents to pay for additional tutoring. There is also the concept of opportunity costs. Children who are in school are unable to produce income working on the family farm or selling in the marketplace. Therefore, when schools do not provide adequate learning, or even basic literacy and numeracy, parents choose not to continue to invest in their children's education.

Sadly, there is evidence that learning outcomes have not increased at the levels expected in many destinations. Learning stagnation is frequently linked to educational institutions within developing countries. This derives from a combination of teacher incentives through to curriculum design and the structure of the education sector in said destination. Predominantly, learning stagnation is a problem associated with political economy, where senior staff involved with the education sector focus on inputs rather than the output of learning. To date, progress in offering free, compulsory and high quality education for all in developing countries has been slow. Overall, women have been the most affected, with limited access to educational opportunities which can largely be attributed to social, traditional and deep-rooted religious and cultural beliefs that are often the barriers they encounter and are unable to overcome.

Teachers are one of the most influential and powerful forces for equity, access and quality in education and are key to sustainable global development. However, their training, recruitment, retention, status and working conditions continue to be problematic areas. While it might sound ideal to have a native English teacher teaching English in a school, particularly if they are paid little, are undertaking voluntary work or are indeed paying for the privilege to teach at the school (although whether the school receives any monetary gain is not always clear – this is discussed further in Chapter Six), their participation in the students' education may have little benefit if they do not possess the right skills and knowledge.

## Ethnocentric Perceptions

One of the inherent difficulties in understanding TEFL tourism is the geographical, cultural and organisational differences across TEFL destinations or language teaching institutions. What one may perceive to be 'good' teaching in one country may not be seen in the same way in another. Class sizes, teaching practices, facilities, bureaucracy, parental involvement, etc. vary significantly across different educational contexts. The problem here arises when teachers or TEFL tourism workers are taken from one context and placed within another.

Whether purposefully or on a subconscious level, it is a common perception that westerners and western practices are superior to those situated in or adopted by developing nations. This view of superiority, otherwise known as ethnocentrism, is prominent across a number of businesses involving international relations and is particularly prominent within the TEFL tourism industry. An ethnocentric person measures other cultures against their own, which they see as the ideal. In the context of TEFL, this is translated into the classroom, where western teaching approaches and practices are generally viewed as the 'best' ways to teach.

Nominalist ontology argues that, by definition, all concepts are formed through subjective interpretations based upon the person's prior knowledge and experience, which in this instance is their experience within various educational systems. Gregory (1970) explains this further through his theory of top-down processing. He argues that perception is a constructive process where cognitive information from past experiences or stored knowledge is used to make inferences about what is perceived. Based on this premise, it can be argued that a stakeholder's perceptions of the TEFL teaching experience are effectively a hypothesis formed from their prior experiences and understanding. For many people, therefore, their perceptions are likely to be rooted according to practices in their own countries or cultures and the educational institutions within which they have studied or worked.

The word 'ethnocentrism' has its roots in the Greek word *ethnos*, which means 'nation' or 'people', and the English word 'centre'. A common idiom for ethnocentrism is 'tunnel vision'. In contrast, cultural relativism is the belief that individual views and value systems are culturally relative. No single ethnic group has the right to impose their particular system of beliefs and values on another, nor is their worldview superior to anyone else's system of beliefs and values. In effect, what might be right for one culture could be wrong for another. While TEFL tourists may believe this to be their way of thinking, there is

**Box 5.3.** Ethnocentrism in TEFL

Deveney's (2005) research demonstrates that foreign teachers in Thailand do not always perceive the cultural aspects of teaching in Thailand prior to commencing employment accurately, which consequently is a major source of frustration to the teacher who is used to working within the cultural norms of western society. The three fundamental discourses of self, social harmony and *sanuk* (fun/merriment) are so deeply ingrained within Thai culture that the teacher must significantly adapt their practices, not according to their previous knowledge and experiences based on what they have learned in their home country, but instead according to local customs, beliefs and practices. As a result, the foreign TEFL teacher may be ill-equipped for dealing with issues such as inherent shyness, lack of parental intervention, hypersensitivity to criticism or excessive cheating.

always going to be the inevitable underlying philosophy of top-down processing (Gregory, 1970), which impacts the tourist's actions and thoughts, whether this be intentional or not.

Ethnocentrism within education occurs when the curriculum, teaching methods and educational strategies used impose the history and cultural aspects of a nation from a monocultural western European perspective, which ignores the presence and/or contributions of other groups in society. This can be seen, for example, when racial/ethnic minorities are absent in the educational literature used in the classroom and/or members of these groups are portrayed in stereotypical and negative ways. While the presence of ethnocentrism may be negative in the most part, failure to account for individualism and cultural differences through the view that the western system is indeed superior to the system in question in some instances can produce positive outcomes.

It is clear that developing countries are not achieving the educational outcomes that western countries do. While, of course, financial difficulties, political infrastructure and the local economy play an inevitable role in the success of education, there is an argument to say that the West is doing it 'right' and the developing world are doing it 'wrong'. There is a strong view that if developing countries adopt some of the approaches and practices utilised in western nations then their educational outcomes are likely to be improved. This very notion does, however, imply the 'superiority' that is deeply infused within the concept of ethnocentrism.

In British schools, the curriculum has been described as ethnocentric, because it gives priority to white culture and the English language (Troyna and Williams, 1986). While nowadays the curriculum is a lot more inclusive of other cultures and nationalities, this highlights that the issue is not only encountered in developing countries. Coard (2005) explains how the ethnocentric curriculum may produce underachievement. For example, in history the British are presented as bringing civilisation to the 'primitive' peoples they colonised. This image of black people as inferior undermines black children's self-esteem and potentially contributes to reduced academic outcomes.

This highlights that the educational system and the curriculum that a destination or institution adopts should not be narrow-minded in approach, but be inclusive. Exposure to different cultures and ways of life is important, but this should not be to the detriment of local priorities. So, for example, it might be useful to teach students in rural Uganda what the stock market is, but priority is better placed on managing personal finances and small businesses, rather than the financial district in London.

Western educational systems are more developed than those traditionally adopted in developing countries, but that does not mean that the best approach is to simply copy what is done in the West and place it in another destination. While some elements may well be superior to what is done in the developing world, local practices should not be undermined. It is clear here that contextualisation is key and this can only be achieved if those employed in the education system are familiar with local needs and customs. Good practice identified in the West can be noted and adapted to suit the given situation. This is where working with local stakeholders is important. It is therefore imperative that TEFL tourists should work closely with members of the local community to ensure that the provision taught is most suitable for their needs.

## Pedagogy and Teaching Practice

Pre-conception of what TEFL involves is not only limited to culturally-infused perceptions, but also prescribed teaching standards. Although often little is known about the TEFL teachers themselves – their characteristics, motivations and everyday practices – there is abundant information on the educational systems from which their ethnocentric perceptions of TEFL may derive. When searching for TEFL opportunities on the internet, it can be seen that many organisations specify that only a few nationalities are accepted onto their programmes, frequently the UK, USA, Ireland, Australia/New Zealand and less commonly South Africa. Based on the above argument, therefore, it can be presupposed that many stakeholders' perceptions are likely to be grounded in the ideals and practices in the these nations.

In contrast to the limited literature regarding teaching standards in many developing countries, it is widely recognised in western countries, including those named above, that teaching quality is the single most important factor affecting student learning outcomes. Consequentially, many nations have prescribed standards and expectations in place in order to define what is expected of teachers. The Centre of Study for Policies and Practices in Education (CEPPE) (2013) presupposes four objectives for defining standards for teaching: to support the improvement of teacher performance; certification of teachers who are new to the profession; assessment of performance; and evaluation and accreditation of teacher-training institutions. Other national standards for teachers often address similar content across different formats, ranging from specific to generic. Table 5.1 provides an indication of the typical ethnocentric expectations of English teaching held by TEFL stakeholders based on published standards, identifying a range of concepts encapsulated within the holistic teaching approach in the named nations, which are by no means exhaustive. This is in stark contrast to the teaching expectations in many developing countries, which are considerably more vague and less prescriptive.

While pedagogically these are important issues, it is not the intention of this book to examine teaching practices per se, but instead to provide indication of the extent to which TEFL teaching may be representative of teaching in the TEFL teacher's home country as a career choice or whether it more closely resembles a form of tourism. It is also important to note that this data is based on available literature at the time of research, which was limited in part, particularly regarding teaching standards in Ireland and South Africa where the focus appears to be on ethical practice rather than pedagogical practice. Furthermore, teaching standards in different nations are not always directly comparable. As such, the data provided is indicative as opposed to descriptive, accounting only for explicit evidence available for each of the presented expectations.

Interpretation of teaching standards can be deemed problematic as a result of their commonly subjective nature, thus contributing to the lack of clarity when defining TEFL teaching. This subjectivity is particularly relevant when considering the global nature of the TEFL industry, where aspects may be perceived differently in different parts of the world. There is also the issue of monitoring and regulation – it is extremely difficult to measure whether all teachers are meeting the required standards at all times. The lack of current regulation of professional standards within TEFL is a particular concern, with different countries having their own approaches regarding standardisation, some with heavier regulatory requirements and enforcements than others.

Not only do standards differ considerably between TEFL destinations, causing a level of confusion for the TEFL teacher, but teaching practices vary significantly also. Take behaviour management, for example. In the UK, depending on the school, this can be a dominant aspect of teaching, where sanctions are regularly handed out to unruly students. In other parts of the world, this may be completely different. In South-East Asia, for example, children tend to be very respectful and rarely misbehave.

Every educational institution has its own way of dealing with bad behaviour, whether this be via a formal behaviour policy or simply through a 'way of doing things'. In the UK, punishments tend to centralise around sanctions which involve admonishing the student, sending letters home to parents, removal of said student from the situation, class or group, confiscating items, detention, etc. Many TEFL teachers will be familiar with punishments of this type given that they were probably educated in a system which operated in a similar way. Difficulty can arise, however, when the TEFL teacher is placed in a system which operates differently and where cultural norms significantly differ from those that the teacher is used to.

One of the most notable differences that affects many TEFL teachers psychologically is that of physical contact. It is hard to believe that just 30 years ago in the UK, teachers used to routinely hit children who did not behave as expected. This was permitted in the context of acting in loco parentis, which meant that teachers were seen as authority figures and were therefore granted the same rights as parents. Physical punishments tended to derive from the use of a cane or a slipper which were used to smack children on their hands or buttocks, sometimes repeatedly, although teachers were free to use whatever punishment they saw fit.

**Table 5.1.** Typical teaching standards in western nations

| Area | Expectation | UK | Ireland | USA | Aus ralia | South Africa |
|---|---|---|---|---|---|---|
| Inspiration, motivation and challenging environments | Establish a safe and stimulating environment | X | | | X | |
| | Enable development of lifelong learning skills | | X | X | | |
| | Set goals that stretch and challenge | X | X | X | X | |
| | Demonstrate and encourage positive attitudes, values and behaviour | X | | | | |
| | Encourage critical thinking, creativity and problem-solving | | | X | | |
| Progress and outcomes for students | Maintain high standards for attainment, progress and outcomes | X | X | X | | |
| | Acknowledge capabilities and prior knowledge when planning | X | | X | | |
| | Encourage reflection of progress | X | | | | |
| | Encourage pupils to take a responsible and conscientious attitude to work and study | X | | | | |
| | Demonstrate understanding of how learning takes place | X | | X | X | |
| Subject and curriculum knowledge | Demonstrate sound knowledge of the relevant subject(s) and curriculum areas | X | | X | X | |
| | Promote high standards of communication, literacy, articulacy and correct use of English language | X | | X | X | |
| | Understand and link curriculum to current interdisciplinary themes | | | X | | |
| | Demonstrate a clear understanding of appropriate teaching strategies | X | | X | X | |
| Planning and teaching | Impart knowledge and develop understanding through effective lesson planning | X | X | X | X | |
| | Promote a love of learning and intellectual curiosity | X | | | | |
| | Set homework and out-of-class activities to consolidate and extend knowledge | X | | | | |
| | Promote the use of interactive technologies in lessons | | | X | X | |
| | Use a variety of teaching strategies | | | X | X | |
| | Contribute to the design and provision of an effective curriculum | X | | X | | |
| Teaching adaptations/ flexibility | Differentiate appropriately according to the individuality of students and their needs | X | X | X | X | |
| | Understand factors that may inhibit learning and how to overcome these | X | | | | |
| | Demonstrate an awareness of physical, social and intellectual development | X | | X | X | |
| | Understand the needs of students with special educational needs, high ability, English as a second language and disabilities, adapting teaching where appropriate | X | | X | X | |
| | Understanding the role of aspects such as language, culture, religion and socioeconomic status in learning, adapting teaching where appropriate | | | X | X | X |
| Assessment | Undertake relevant assessment, including statutory assessment requirements | X | | X | X | |
| | Incorporate and design a range of assessment types | | | X | | |
| | Make use of formative and summative assessment | X | | X | X | |
| | Use relevant data to monitor progress and adapt processes where necessary | X | | X | X | |
| | Give regular feedback and report progress | X | | X | X | |

| | | | | | | |
|---|---|---|---|---|---|---|
| Behaviour management | Maintain clear rules and routines and promote good behaviour | X | | | X | |
| | Establish a framework for discipline with a range of strategies | X | | | X | |
| | Refrain from physical or psychological abuse and avoid inappropriate physical contact | | | | | X |
| | Exercise authority with compassion | X | | | | X |
| Wider professional responsibilities | Make a positive contribution to the wider life and ethos of the school | X | | | | |
| | Develop effective professional relationships with colleagues | X | | | | X |
| | Deploy support staff effectively | X | | | | |
| | Report incidents or matters which impact on student welfare | | X | | | |
| | Communicate effectively with parents and associated stakeholders | X | X | X | X | X |
| Professional integrity and standards of ethics | Treat students with dignity, respect privacy and at all times observe appropriate boundaries | X | X | | | X |
| | Take all reasonable steps to safeguard student well-being | X | X | | | X |
| | Show tolerance of and respect for the rights of others | X | X | | | |
| | Act with honesty and integrity | | X | | | |
| | Do not undermine fundamental national values, including democracy, the rule of law, individual liberty and mutual respect | X | | X | | X |
| | Be tolerant of those with different faiths and beliefs | X | | X | | |
| | Have an understanding of, and always act within, the statutory frameworks | X | X | X | X | |
| | Avoid conflict between professional work and private interests | | X | | | |
| | Do not exploit student vulnerability or activities which might lead them to break the law | X | | | | X |
| | Do not teach while under the influence of any substance which impairs ability | | X | | | |
| | Avoid sexual relations with students | | | | | X |
| | Strive to enable students to develop a set of values consistent with the rights set out by the constitution | | | | | X |
| Professional values and relationships | Maintain professional regard for the ethos, policies and practices of the school | X | X | X | | X |
| | Be caring, fair and committed to the best interests of the students | | X | X | | |
| | Be committed to equality, inclusion and respect by accommodating diversity in gender, civil status, family status, sexual orientation, religion, age, disability, race, ethnicity, etc. | | X | | X | |
| | Seek to develop positive relationships and work collaboratively with students, colleagues, parents, school management and others in the school community | | X | X | X | X |
| | Work to establish and maintain a culture of mutual trust and respect | | X | | | |
| | Respect the authority and positions of colleagues | | | | | X |
| Continuous professional development | Be open and responsive to constructive feedback and seek support if necessary | | X | | X | |
| | Maintain professional knowledge and understanding to ensure it is current | X | X | X | | X |
| | Reflect on and critically evaluate professional practice | X | X | X | X | |
| | Seek opportunities for career-long professional development | X | X | X | | X |
| | Identify and seek learning opportunities | | | X | X | |
| | Participate in professional and community networks | | | | X | |
| | Support new members of the teaching profession | | | | | X |

**Box 5.4.** TEFL overseas is not preparatory for life in a UK classroom

Working in a school where there is no formal curriculum, no standardised assessments, where playing games is the norm and students are impeccably behaved, is not preparatory for a teaching career in the UK. Yet increasingly young people are choosing to spend their gap years teaching English as a foreign language (TEFL) overseas – in exotic destinations such as Thailand, China or Costa Rica, where they believe they are gaining invaluable teaching experience to take back to a UK classroom.

'Teacher training' is commonly provided by companies which resemble a tour operator more closely than an education provider. Participants pay an often extortionate fee for a few days training alongside leisure time. In contrast, a PGCE in the UK requires extensive evaluations of pedagogical literature, countless teaching observations and teaching experience equivalent to a full-time teaching position. This process is a long and arduous one and is not comparable to the majority of TEFL teacher training courses offered by such operators.

Once 'qualified', daily TEFL activities frequently consist of playing games with the students, designing and writing exams and taking part in social events, such as a school play or a teacher gathering. This [is] by no means representative of teaching in the UK, although many TEFL teachers do not realise this.

Securing a place on a PGCE programme in the UK at a good university is no easy feat and many people undertake TEFL placements in order to gain experience and to enhance their CV. Little do they know, TEFL is a walk in the park compared to what they will experience back home. Differentiation. Plenaries. Behaviour management. Schemes of work. Administrative duties. These are concepts that many TEFL teachers have no idea about. They are ill-prepared for a teaching career in the UK and disillusioned about what teaching really is.

This is not only an issue for the TEFL teacher. The universities who recruit them and the schools who hire them are generally not aware of these stark differences in educational systems and job profiles, resulting in employers and educational providers having unrealistic expectations of the skills of the teacher whom they have recently recruited.

(Stainton, 2017a)

The ban on corporal punishment came into force in 1986 in British state schools (private schools in 1998 in England and Wales, 2000 in Scotland and 2003 in Northern Ireland) and a similar pattern is evident across most of the western world. Those TEFL teachers who began their schooling after this time are unfamiliar with the concept of physical punishment being used in schools. Following on from this, a culture has developed whereby physical punishment is largely frowned upon, not only in an educational setting, but also in the home, with a new law coming into force in 2004 stating that it is illegal for a parent or carer to smack their own child, except where it amounts to 'reasonable punishment'.

Given the general perception, therefore, that hitting children as a means of punishment in an educational setting is unacceptable, it can be quite a shock to TEFL teachers when they encounter it abroad. There are 67 countries in which corporal punishment is still legal and many countries, while they have laws against the physical punishment of children in schools, do not strictly enforce this. In many parts of the world, it remains culturally acceptable to use physical punishment to enforce behaviour management and this is something that many TEFL teachers struggle to come to terms with.

The United Nations has developed an educational model that combines traditional teaching content with life skills covering finances, health and administrative skills, which are delivered via existing school systems and teachers. Referred to as 'school for life', this model shifts the goal of schooling away from the achievement of standardised learning outcomes towards making a positive impact on the economic and social well-being of students and their host communities. For western parties involved, this means suspending their own beliefs and ethnocentric perceptions to involve locals in the decision-making processes and curriculum design and to ensure that it suitably addresses the needs of the community.

**Box 5.5.** Behaviour management

Societal, legal and ethical perspectives have all contributed to the modern approach to behaviour management seen in developed countries today. In support of the individualist notion evident throughout contemporary western educational systems, schools and educational institutions now operate on a more inclusive basis than in the past (Wearmouth and Glynn, 2013), often utilising positive disciplinary approaches such as guiding and teaching, encouraging the student to understand the consequences of their actions, teaching students to problem-solve and develop self-image, self-control and cooperation (Sailor, 2010).

Rather than adopting long-term strategies for behavioural management, however, in Thailand teachers traditionally use aversive control that can cause physical pain. Examples of such practices include canning, scolding, flogging and pinching (Pumwaree, 1986). In an attempt to minimise these unethical practices, canning was banned and there was a call for alternative methods more in line with western practices, such as assigning punitive activities, probation or suspension (Walker *et al.*, 2004). Physical punishment of children in Thailand, however, remains socially acceptable and is commonly practised among many Thai families as well as educational institutions despite the existence of the ban. This was evidenced by Jampian (2012) who demonstrated through his research that 72.5% of children were subjected to corporal punishment in 2011.

This requires significant changes in terms of both content and pedagogy within the educational institutions of developing countries around the world. Entrepreneurship and health modules are suggested as mandatory curriculum components for all primary grade students, as these are identified as essential areas in need of development. The entrepreneurship curriculum is informed by work with adult entrepreneurs in developing countries, drawing upon ideas from a range of financial and entrepreneurial programmes developed by known organisations. Health modules focus on preventing disease, caring for sick children and obtaining medical care. Classes are designed to encourage student-centred learning methods that require students to work in groups to solve complex problems and manage projects on their own. This is in contrast to the typical rote learning approach that is utilised in many countries.

The traditional outcomes associated with 'school quality' in the developing world are based largely on the use of traditional schooling approaches (e.g. good test scores). However, it is clear that these do not produce the best outcomes in terms of the students' livelihoods. There is a changing dynamic where interventions in educational systems are leaning towards the greatest social and economic impact for the poor as opposed to the highest test results. While this is likely to be implemented largely from a top-down perspective, TEFL tourists do play an important part. Teaching language that is most useful to the local community in a way that they are likely to understand and relate to best is imperative. In order to achieve this, however, sufficient training must be provided for the TEFL teacher. This is discussed further in Chapter Six.

Aside from standards and curriculum design, there are other elements integral to teaching practice which differ significantly according to geographical and cultural contexts. One such area is workloads and working duties. There has been considerable controversy in recent years regarding the workload of teachers in developed countries. While there is evidence in the media of this to an extent in all TEFL teacher-originating countries, the problem appears especially prominent in the UK. The OECD found in its Teaching and Learning International Survey (TALIS) (2014) that the international average number of contact teaching hours per week was 19.6. However, after accounting for other tasks such as marking, planning and administration, British teachers worked a total of 46 hours per week, 8 hours above the international average. The study also found that only 35% of teachers in developed countries felt valued compared to two-thirds of teachers in South-East Asian countries.

Pells (2015) writes that the strain of workload causes up to 4 in 10 new teachers to resign in the UK within a year of qualifying, largely attributed to the political bureaucracy, arbitrary administration duties, financial pressures, Ofsted requirements and pressure to achieve particular grades. The story is, however, quite different in developing nations. Said countries

**Box 5.6.** Teachers moving overseas

Many British teachers are leaving the UK state education system and many more are planning to do so. Many of these are switching the UK classroom for an international school.

The reasons behind so many people leaving the profession vary. According to recent research compiled by A* Star Teachers, the prime causes seem to be the excessive workload, the need for a better work/life balance, unreasonable demands and too much organisational change.

In the first half of 2016, the sector experienced a loss of 9.5% full-time staff, says the study – 57% of the teachers who were surveyed said they were leaving to seek a better work/life balance, a situation often quoted by teachers who have moved to international schools.

Many international schools provide employment packages far more competitive than that offered in the UK, are free from the arbitrary bureaucracy of Ofsted and enable teachers to have a better work/life balance.

(TIC Recruitment, 2018)

typically have less administration requirements, have less paperwork to complete and have fewer formal regulatory inspections in the way that Ofsted provide. While this can be viewed as a positive aspect from the perspective of the teacher working in the developing country, it can be the cause of frustrations, particularly when elements are not organised or managed as efficiently and effectively as the TEFL tourist may be used to. Furthermore, such experiences can be deemed not to prepare teachers sufficiently for life in the UK classroom, as noted in Box 5.4.

## The Definition of 'Teacher'

In light of the above discussions regarding teacher duties and the differences between the role of a teacher in the UK compared with many developing nations, there is a natural urge to ask the question: 'Is a TEFL tourist actually a teacher?' This question is inherently subjective and perhaps controversial as well. A TEFL teacher may well believe that they are professional as they have a TEFL qualification and they work in a school on a full-time basis. A western qualified teacher who has spent years training in the profession may disagree.

Informally the role of a teacher can be taken on by anyone (e.g. when showing a colleague how to perform a specific task). In some countries, teaching young people of school age may be carried out in an informal setting, such as within the family by adopting a home-schooling approach rather than in a formal setting such as a school or other educational institution. Some professions also involve a significant amount of teaching, for example a youth worker or a pastor. In many parts of the world, formal teaching of students is usually carried out by paid professional teachers. The problem is, however, defining a 'professional' teacher.

Essentially, teachers are there to facilitate student learning, often in a school or academy or perhaps in another environment such as outdoors. The primary objective is to use either an informal or formal approach to learning to enable knowledge development and thinking skills. There are a range of different pedagogic approaches to learning and those who have been sufficiently trained will have the skills to differentiate their teaching practice according to students' background knowledge, environment and learning goals. They will also be familiar with the relevant curricula as determined by the local authority in the locality within which they are working. Teachers may also assist in learning outside the classroom by organising student trips or enrichment activities. The increasing use of technology in the classroom environment has also opened up new doors for teachers.

Teaching using pedagogy also involves assessing the educational levels of the students in reference to their English language, or other relevant abilities. Understanding the pedagogy of the students in a classroom involves using differentiated instruction as well as supervision to meet the needs of all students in the classroom and it is unlikely that a TEFL teacher will have adequate knowledge in this regard unless they have had formal educational training. This is really one of the fundamental problems with TEFL tourism: that the tourists most commonly have had little more than a short course of training, whereby the teacher trainers are usually no more qualified than the tourist is upon completion. This general lack of sufficient training can have far-reaching consequences for both the learner and the society within which the TEFL tourism takes place. This is addressed further later in this chapter.

There is also the issue of enhanced globalisation in tourist-receiving areas. In response to the postmodern globalisation era characterised by increasingly non-restrictive geographical boundaries, many educational institutions across the world have developed strategies for internationalising their educational programmes. In many instances this has been undertaken through 'copying' the West, with educational reforms strongly influenced by western doctrines, ideologies and practices. This does not, however, facilitate easy comparability between educational systems and experiences.

The ambiguities in defining the term 'international education' have been emphasised by Hayden (2006), who explains that the term is used within three dominant spheres: teaching international-based subjects such as development; the integration of educational systems beyond national boundaries; and the comparison of educational practices between different countries. The United Nations Educational, Scientific and Cultural Organisation (UNESCO) on the other hand, understands international education as a process resulting from international understanding, cooperation and peace. In the most part, this book addresses international education through Hayden's (2006) third lens by addressing the practice of TEFL across different institution types in different localities. However, it is also important to acknowledge the complexity of international understanding.

This can be demonstrated, for example, through the subjective notion of what constitutes a teacher and, on a more complex level, a 'good' teacher. In the broadest sense, a teacher is an educator, a facilitator of knowledge transition from one person to another, but defining an effective teacher permeates beyond this (Clark, 1993). There have been a number of attempts at defining effective teaching. However the majority take a generic approach, thus failing to take into account that individual classrooms within different cultures can vary greatly in terms of the expectations of teachers and students. This therefore indicates that what one may perceive as being 'good' teaching in one country may not be classified as 'good' in another.

To emphasise this point, we can draw upon a selection of perspectives that exist among the huge array of pedagogical literature. Vogt (1984), for example, highlights the importance of differentiation between students of different abilities, stating that teachers should provide students with clear learning objectives, while continually assessing individual students' progress. Collins (1990) suggests that effective teachers should be committed to students and learning, have good subject knowledge, be able to manage students responsibly, be reflective and systematic in their teaching approach and be active members of the learning community. Wenglinsky (2000) ascertains that practices promoting higher-order thinking and active participation are most successful. Swank *et al.* (1989) developed a model of effectiveness based upon teacher actions, where it is suggested that teachers should increase academic questions and decrease lecture and negative feedback. Papanastasiou (1999) argues that no single teacher attribute or characteristic is sufficient in defining an effective teacher.

Taking this argument one step further, some researchers suggest that the effectiveness of teaching is best measured not according to the teacher, but to their student achievement (e.g. Clark, 1993; Sanders and Rivers, 1996; Wenglinsky, 2000). Measuring student gains, however, brings yet another subjective dimension to the debate. Goldstein (2001), for example, asserts that it is very difficult to attribute the progress of any one student in a given subject entirely to the teacher of that subject, given that there is a plethora of other factors that influence the student's ability to learn effectively, such as background, other teachers and school setting. It would therefore be unsuitable to attribute effective learning only to the teaching that has been provided. Further, Long (2000) concludes that there is no established connection between learning and teaching, and that there is little variation between teachers in terms of their impact on students' progress.

It is evident from the academic debates presented here that the notion of good or effective teaching is subjective and that these subjectivities are exemplified when added to the mix of cultural and geographical factors to which the international education sector is subjected.

## TEFL Training and Qualifications

There is an abundance of qualifications available to prospective TEFL teachers at vastly different levels and costs, ranging from short introductory courses to distance learning and academic programmes. These cover the range of acronyms explained in Chapter Two, including TEFL, TESOL, ELT and ESL. In practical terms, there is little or no difference between a TEFL and TESOL qualification. They are both acronyms that essentially mean the same thing – teaching the English language to non-native speakers – and are used interchangeably in the industry. TESOL (Teachers of English to Speakers of Other Languages) is the term which refers to the teaching of English to non-native speakers who are based in native English speaking countries. People who study and work in the UK, Canada, Australia, Ireland, New Zealand, South Africa and the USA will typically use the term TESOL. While tourists can undertake TESOL work in this regard, this is more likely to take the form of a formal career choice as opposed to a temporary employment and travel opportunity.

There is no requirement to possess a degree in education, have prior teaching experience or hold a university degree to get paid to teach English abroad. Typically, to teach English as a foreign language in a non-native English-speaking country, some form of English language certification is required. However, there is considerable ambiguity. Qualifications can be undertaken prior to departure (e.g. the Cambridge Certificate in English Language Teaching to Adults (CELTA) course) (Box 5.7), or in an educational institution in the TEFL destination or as part of a 'package style' programme as offered by organisations such as Experience Teaching Abroad. Such packages frequently offer a combined TEFL qualification and in-country training with a guaranteed job, accommodation and assistance in organising flights.

There is a broad range of bodies certified to deliver TEFL qualifications, differing in size, structure and nature, with little apparent differentiation between the TEFL certificates awarded.

**Box 5.7.** Cambridge Certificate in English Language Teaching to Adults (CELTA)

CELTA is one of the most well known qualifications in the TEFL industry. It is a teacher training qualification provided by Cambridge English Language Assessment through authorised Cambridge English Teaching Qualification centres and can be taken either full-time or part-time. The qualification was developed to be suitable both for those interested in Teaching English as a Foreign Language (TEFL) and for Teaching English to the Speakers of Other Languages (TESOL).

TEFL certification originated in International House, London, in 1962 and was run from the 1970s until 1988 by the Royal Society of Arts (RSA). Then in 1988, Cambridge English Language Assessment (known as the University of Cambridge Local Exam Syndicate (UCLES)) took over the running of training courses for TEFL teachers. The qualification was developed in response to perceived student needs, leading to the launch in September 1988 of the CTEFLA (the Certificate in Teaching English as a Foreign Language to Adults). By 1996, the CTEFLA became the RSA/Cambridge CELTA and in 2001 it became known as Cambridge CELTA.

The CELTA course enables students to develop subject knowledge and a working knowledge of the principles of effective teaching. It also enables them to acquire a variety of practical teaching skills for use with adult learners. Students are then provided with the opportunity to demonstrate their ability to apply their learning in a real world teaching environment.

The CELTA syllabus consists of five topic areas:

- Topic 1 – Learners and teachers, and the teaching and learning context
- Topic 2 – Language analysis and awareness
- Topic 3 – Language skills: reading, listening, speaking and writing
- Topic 4 – Planning and resources for different teaching contexts
- Topic 5 – Developing teaching skills and professionalism

All courses have a minimum of 120 contact teaching hours which includes input, supervised lesson planning, teaching practice, feedback on teaching, peer observation, observation of experienced teachers and consultation time. In addition, students are expected to spend an additional 80 hours reading, researching and preparing assignments and lessons.

(Cambridge English Assessment, 2018)

**Box 5.8.** The TEFL 'package'– i-to-i

i-to-i is a subsidiary of TUI, a profit-maximising tour operator specialising in the package holiday model. It is no surprise therefore that there is a distinct resemblance of the traditional package holiday structure within its TEFL programmes. i-to-i is independently regulated by Ofqual and has accreditation from the Open and Distance Learning Quality Council (ODLQC). i-to-i uses the UK qualifications framework which is recognised and replicated around the world. The i-to-i TEFL courses lead to a Level 3 qualification which is the equivalent in difficulty to an A level in the UK, a Collegeboard Advanced Placement exam (AP) in the USA, a Senior Secondary Certificate of Education (SSCE) in Australia or a Matric Exam in South Africa.

A typical TEFL package provided by i-to-i will include the following:

- 120 hours TEFL training
- advice and pre-departure preparation
- visa support
- airport pick-up and transfers
- orientation week and meals
- tour of the local area
- language lessons
- accommodation
- a living allowance
- bonus payment upon completion of placement
- ongoing in-country support
- academic advice in the future
- certificate
- references

(i-to-i, 2017)

A simple internet search for the terms 'TEFL certification' or 'teaching English abroad' will quickly demonstrate that there is a myriad of different TEFL training programmes and courses for the prospective teacher (Fig. 5.1). Given the global nature of the TEFL industry and the varying qualifications offered in different countries, the value of TEFL qualifications is both ambiguous and questionable. Many TEFL stakeholders would be unlikely to be able to differentiate between a highly respected qualification that incorporates substantial teaching practice and a number of assessments such as a CELTA, or an online course that can be completed in as little as 40 hours. It can thus be argued that clarity and regulation of such qualifications is imperative.

TEFL courses generally take one of four approaches. The more formal and well-recognised programmes tend to be classroom-based such as CELTA. This is the most common type of TEFL course and can range from day courses to full-time degree programmes. The more extensive courses will usually incorporate a practicum. However, a substantial classroom-based element does not prescribe that the course is well recognised.

An alternative is the work-based learning approach. This is infrequently used for TEFL certificates, although many might argue that it should be used more. Work-based learning is an educational strategy that provides students with real-life work experiences where they can apply academic and technical skills, and is increasingly used nowadays in training situations, particularly within the education industry. In the UK, there is a current drive towards this training approach and the Department for Education (2017) encourages prospective applicants to apply for 'school-led' teacher training programmes. Despite its popularity in education, this is not currently represented within the TEFL training sector.

Online TEFL training is less formal than classroom or workplace-based courses and is closely aligned with the distance learning approach commonly adopted by educational institutions across the world. Distance learning provides a way of learning remotely without

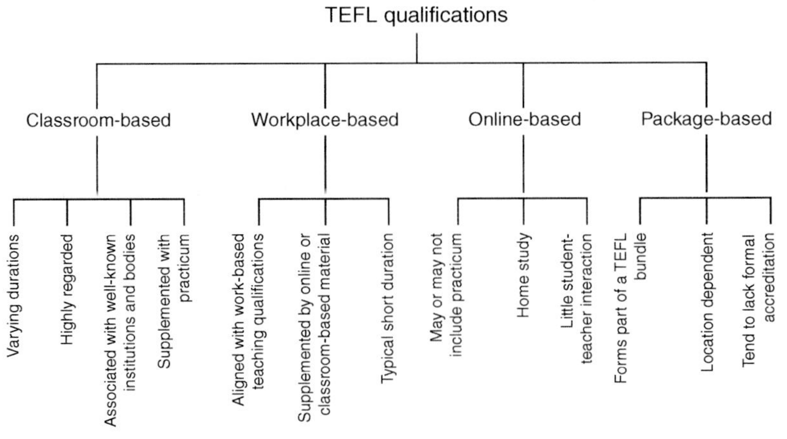


**Fig. 5.1.** Types of TEFL qualifications

being in regular face-to-face contact with a teacher in the classroom and it has experienced exponential growth in recent years as a result of the advent of the internet and widespread use of the computer. This method of study is becoming so popular that, according to the Complete University (2017), there are more than 270,000 undergraduate students and 108,000 postgraduate students studying for their degrees via the internet. Online TEFL courses are widespread and vary considerably in terms of duration, content and accreditation. Courses currently advertised on the internet start at as little as 40 hours upwards.

Package-based TEFL training approaches do not align with the formal education system in the same way. Rather, they sit more closely with the package holiday model, where certain elements are combined to create a TEFL 'bundle'. As the qualification element is part of a larger product, it may not be given the same value and weighting by the TEFL organisation or the TEFL teachers undertaking it. These programmes will often be targeted at younger, more holiday- or travel-focused tourists, and the marketing of said courses invokes notions of adventure and the exotic as opposed to education. These courses may be designed by TEFL agencies who represent tour operators, as opposed to formal training providers, or the TEFL company, which may employ the use of an external TEFL training provider to facilitate this element of the package.

Each type of TEFL course has its merits and limitations. Online courses help to overcome geographical boundaries, but lack opportunity for the student to ask for clarification or additional guidance (a lot of tutorials in education are online through Skype, Facetime, etc.). Meanwhile, work-based learning provides students with work experience opportunities, but they may be exposed to situations for which they are ill-prepared. TEFL packages provide the security of a bundle-style product, but may not provide the educational formalities accompanying a classroom-based course. The issue lies with the inability to easily compare TEFL courses and for stakeholders to understand the difference between, say, an online programme which has been undertaken in as little as 40 hours and a well-recognised CELTA course which has required considerable input in terms of both academic study and practicum.

One way to overcome this is through the use of accreditation. Accreditation ensures that certification courses meet quality standards in the level of instruction, the quality of the curriculum, inclusion of a practicum, extent of training and other areas. It ensures that quality and ethical standards are maintained and recognises professional acceptance in the international job market. Accreditation involves a formal and rigorous process of review by a recognised accrediting body. This is generally an independent third party that examines an educational institution's academic and professional standards. While this sounds like the perfect solution to overcome the ambiguous TEFL qualification market, the term 'accreditation' is loosely applied

**Box 5.9.** ACCET accreditation

The Accrediting Council for Continuing Education & Training (ACCET) is a voluntary group of educational organisations affiliated for the purpose of improving continuing education and training. Through its support of an independent Accrediting Commission, ACCET membership ensures the Standards for Accreditation along with policies and procedures that measure and ensure educational standards of quality.

The fundamental purpose of the accreditation process is to provide a professional judgment as to the quality of the educational institution and programme(s) offered, and to encourage continual improvement thereof. Accreditation is the process by which the ACCET Accrediting Commission grants public recognition to an educational institution which has voluntarily submitted to an evaluative comparison with established qualifications and standards, both upon initial application and periodically thereafter. ACCET's function is focused upon educational quality, the continuous assessment and improvement of educational practices, and assistance to member institutions in the improvement of these practices within their programmes.

The four major components of this on-going process are: (1) the development and implementation of institutional and programme goals and objectives that focus upon a continuing education mission, (2) a comprehensive, analytic self-evaluation review and report by the institution, (3) an on-site professional peer review to evaluate the adequacy and accuracy of the self-evaluation, and (4) an independent review and decision by the Accrediting Commission as an assessment of the institution's commitment to and compliance with ACCET standards, policies and procedures.

(ACCET, 2018)

as there is no one single organisation that is universally regarded as the official accrediting body for TEFL or TESOL certifications. Well-known accrediting bodies include: TQUK, College of Teachers and ACCET. Courses offered directly through an accredited university will also be considered accredited, including SIT, Cambridge University (CELTA) and Trinity University (Trinity certificate).

While accreditation may be a step in the right direction in terms of uniformity and regulation of the TEFL industry, this is ambiguous at most. Not only is accreditation optional for TEFL organisations, but the very fact that there is no one set accrediting body makes the nature of accreditation difficult for the consumer to understand. To add to this perplexity, some educational institutions without a true accreditation have set up their own 'accrediting body' to provide them with a faux accreditation that is not actually recognised in the TEFL field. Said organisations may appear to have official accreditation boasting the likes of a website and an impressive-sounding name, but in fact they do not. Not only is this confusing for the consumer, but it is also unethical.

To further add to this confusion, some TEFL organisations do not obtain accreditation from a recognised (or otherwise) body, but instead become members of various associations. Memberships of academic associations are not based on any criteria and should not be considered the equivalent or a substitute for accreditation. Most educational institutions can simply join an organisation by paying annual fees, with no independent review of their educational and professional standards. These are basically the equivalent of subscription-based memberships.

It is common for bona fide accreditation organisations to prescribe that TEFL companies must provide a minimum number of hours (often 100 hours of academic classes and 6–20 hours of live practice teaching to active EFL students). A standard amount of time for a full-time 100-hour course is 20 classroom days or four weeks. Some organisations offer weekend classes of two days or 20-hour courses which provide only one tenth of the above. The same applies to online TEFL certification classes where it is common for the live practicum element to be excluded. Although TEFL certification courses incorporating fewer than 100 hours of training and 6 hours of practicum may not be considered legitimate by educational institutions, there are still many organisations offering TEFL training in smaller capacities, several of which also claim to be 'accredited'.

Another common prerequisite of recognised accreditation organisations is that the instructors

**Box 5.10.** SEE TEFL

In brief, SEE TEFL has the following accreditations, certifications, licensing, and recognition:

**Licensed by the Thai Ministry of Education**

DSS International/SEE TEFL is registered with the Thai Ministry of Education, and their licensed school has permission from the Thai Ministry of Education for our school to conduct TEFL Certification training courses.

**Approved and Endorsed by Training Qualifications UK**

Their most recent endorsement and approval is by (TQUK), the Training Qualifications UK organisation. As an approved centre, courses are endorsed by TQUK, which is recognised by the UK Government Office of Qualifications and Examinations Regulation (OFQUAL).

**Certified ISO 9001:2008 by Bureau Veritas**

SEE TEFL Teacher Training ISO certification was established through Bureau Veritas, an accredited registrar and auditor that performs assessments of management systems against requirements of national and international standards for quality.

ISO Certification by Bureau Veritas is accredited by the United Kingdom Accreditation Service UKAS, the sole national accreditation body recognised by the UK government to assess organisations that provide certification services.

**Thai Industrial Standards Institute Recognition**

The Thai Industrial Standards Institute (TISI) is an office of the Thai Ministry of Industry. One of the objectives of TISI is to monitor the quality of Thai products and services and to promote companies that meet international governmental-level standards such as ISO 9001:2008. SEE TEFL is the only TEFL trainer recognised by TISI as meeting an ISO international standard.

**IATEFL Professional Organisation Membership**

SEE TEFL is an Institutional Member of IATEFL, the International Association of Teachers of English as a Foreign Language, and one of only three institutional members in Thailand.

(SEE TEFL, 2018)

employed to train prospective TEFL teachers are suitably qualified. Typically, a university-level professional training class should be taught by a highly qualified instructor with the equivalent of a masters degree in TESOL or a related field, or equivalent training and experience (i.e. a DELTA or degree in Education, etc.), coupled with significant experience of teaching English as a foreign language. While this may be the case for some of the more formal TEFL courses, the reality is that in many instances TEFL trainers have few qualifications more than those they are teaching and it is common for organisations to recruit past TEFL students with only a few months of practical teaching experience. Said TEFL trainers are often referred to as a 'tutor' or 'mentor' and are recruited based on the fact that they can be offered a salary that is considerably smaller than that required of a university-level instructor.

Many potential employers around the world will not recognise a certification that does not incorporate a practicum and so this is an important criterion for accreditation organisations. While the extent to which qualifications may involve practical teaching experience differs substantially between the organisations offering this, the very nature of practical teaching presents further notions of ambiguity within the TEFL industry. Some international accreditors prescribe that practicum be undertaken with *real* students. This is because some TEFL providers claim that their courses incorporate student teaching, but in reality their courses only incorporate role-playing with other teachers-in-training. Courses incorporating peer-to-peer teaching in this regard are unlikely to prepare the prospective TEFL teacher as well as courses which provide the opportunity for teaching

**Box 5.11.** TEFL training job advertisement

**TEFL Trainer for 8 days at TEFL Heaven (Koh Chang)**

**Job Description**

Echo English is looking for motivated, intelligent and reliable TEFL trainers to train student teachers on our TEFL Heaven program. Training takes place on the paradise island of Koh Chang. Trainees will have just arrived in the country, and your job will be to train them how to teach English.

The job is from Monday 5th August until Tuesday 13th August. During this time there will be 5 days teaching and training of eight newbies to Thailand who will be teaching for one month in Chiang Mai afterwards on a volunteer programme. You also have to attend an induction/ training for you on Friday 3rd August at our office in Bangkok.

During the time your main roles will include leading the training sessions Monday to Friday, observing trainees' lessons and providing the trainees with constructive feedback and marking assignments. The position is fully supported and Echo staff will be alongside you on the island.

Benefits:
Experience in a leadership role
12,000 THB for 6 days of work (1 days training/induction for the trainer by Echo staff, then 5 days training teachers on Koh Chang)
Full accommodation provided (sharing with other staff) on a resort paradise location
Opportunity to enjoy a beautiful resort and to make new friends

We prefer the right candidate to have:
A TEFL Certificate
Experience teaching English in Thailand across a range of ages
Some experience teaching adults in any setting
Someone with strong communication skills and confidence
Someone who is well organised

Prior experience in TEFL training is an advantage but not essential. You also must be reliable, friendly and enthusiastic.

Please send us your CV, a covering letter and a recent smiley picture.

**Location : Koh Chang, Thailand**

**Requirements**

**Required level of English:**
Native speaker
**Certification required:**
A TEFL

(Asia Teaching Jobs, 2018)

practice in a real educational setting and are often adopted in order for the TEFL organisation to save on the money and effort required to organise student placements.

## Teacher Skills and Suitability

The UNWTO's report on *Measuring Employment in the Tourism Industries* (2014) demonstrates the pivotal role that tourism has in providing job opportunities, particularly for low-skilled workers, workers with few qualifications, ethnic minority groups and migrants, unemployed youth and the long-term unemployed. Being a labour-intensive sector, tourism provides the perfect opportunity for employment of persons entering the labour market for the first time or for those having difficulties finding employment elsewhere. This is particularly relevant for the TEFL tourist, who may not necessarily have the skills or suitability to become a teacher in their home country.

One area of the TEFL tourism industry that is especially ambiguous is the criteria to become a teacher. While there are many positions advertised formally which require accredited degree certifications and the appropriate teacher training to be undertaken, others appear to have very few or no formal requirements. This makes it difficult to judge whether a TEFL teacher is indeed a qualified teacher and to what extent they possess the required skillset to provide an adequate education to their TEFL students. In fact, the very term 'qualified' cannot be applied in a global context. In a country where there are no 'formal' requirements prescribed by the government for English language teachers, is the ability to speak English enough to deem a person 'qualified'? This is an ambiguous and somewhat grey area.

**Box 5.12.** Fake TEFL certificates in Bangkok

Many travellers to Bangkok will recognise the stalls selling a range of fake IDs and qualifications. From student cards, driving licences and citizenship cards to degree and TEFL certificates, you can buy almost any fake document in Thailand. The most common place to purchase such documents is Khaosan Road, a haven for backpackers, many of whom may dip their feet into the TEFL cooking pot at some point during their travelling experience. Prices generally range from approximately 250 Bhat upwards and can be produced in as little as two hours. Despite such business being illegal in Thailand, with severe punishment if prosecuted, the authorities often turn a blind eye, giving vendors the opportunity to promote and sell their products in blind sight.

Many studies addressing teacher suitability have avoided the use of the term 'qualified' and instead focused on the aspect of cognition, which is made up of a number of elements as opposed to only the qualifications held. The term 'teacher cognition' refers to the unobservable cognitive dimension of teaching – i.e. what teachers know, believe and think. Teachers are effectively active, thinking decision-makers who make instructional choices by drawing upon complex, personalised, practically-oriented and context-sensitive networks of knowledge, thoughts and beliefs (Borg, 2003). Fig. 5.2 suggests that various psychological

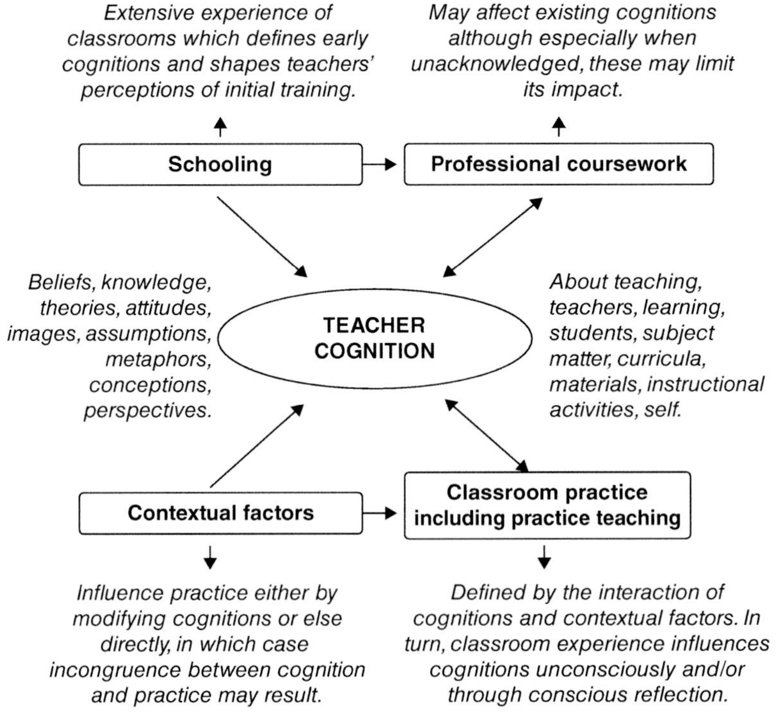


**Fig. 5.2.** Borg's (1997) model of teacher cognition, schooling, professional education and classroom practice

constructs determine a teacher's knowledge and understanding of the environment in which they are working, which then contribute to their ability to be an effective teacher.

Conceptions are 'general mental structures, encompassing beliefs, meanings, concepts, propositions, rules, mental images, preferences and the like' (Brown, 2004, p. 303). A TEFL teacher's conceptions essentially provide frameworks for understanding, interpreting and interacting with the TEFL environment and therefore determine their pedagogic approaches and choices of materials, content and learner activities. To deem whether a TEFL tourist may have the necessary skills to undertake the job, therefore, requires looking at more than simply the qualifications held, and involves taking a rounded view of their conceptions.

Academics have argued that the most influential conceptions derive from early learning and memories of a teacher's own schooling and observations of their own teachers (Mitchell and Weber, 1998; Brown, 2004; Pereira, 2005; Allender and Allender, 2006). This supports the notion of ethnocentrism referred to previously in this chapter, precluding that a TEFL teacher's perceptions of TEFL teaching are inherently based upon their past experiences. Although professional preparation and formal training does play a role in shaping TEFL trainees' cognitions, it is important to recognise that this may not always alter the TEFL teacher's mind-set, with past experience playing such an important role in conception.

The OECD deconstructs this concept further, demonstrating that there are six key elements which affect student learning and student outcomes. Four of these concepts are attributed to the background of the educational institution and the processes that it adopts, while the other two address the beliefs and attitudes of the teacher along with their professional competence. Like Borg, the OECD identify the importance of cognition resulting both from personal experiences, backgrounds and pedagogic knowledge.

The question here then lies with regard to teaching pedagogy and how knowledge of teaching and teaching skills can be measured. Do TEFL teachers have the required cognitive abilities and teaching skills to perform in all TEFL tourism contexts? While there appears to be no data available as yet to support this, it is highly

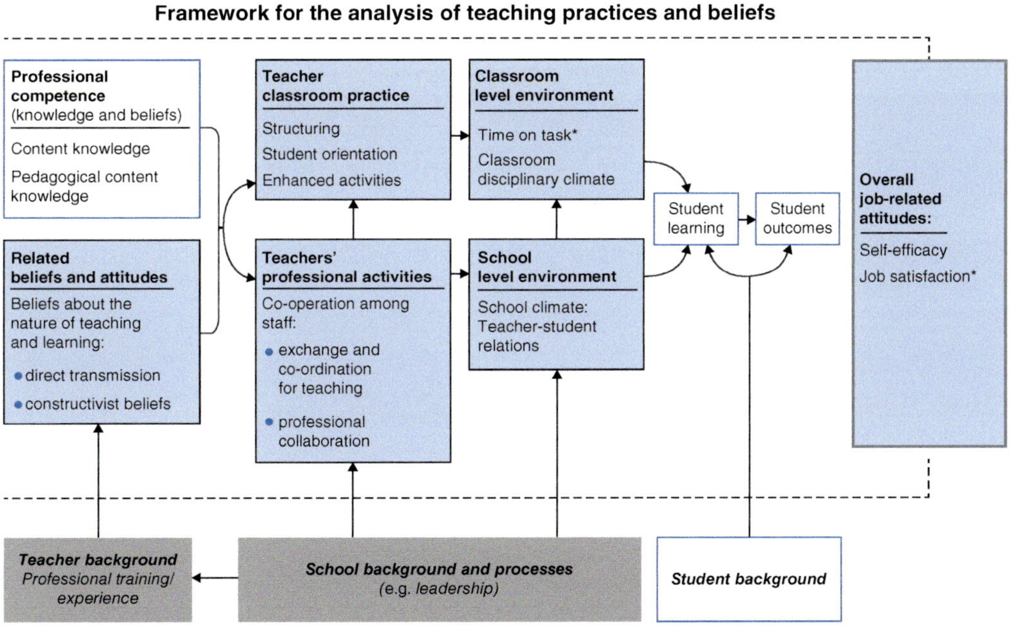


**Fig. 5.3.** OECD's framework for the analysis of teaching practices and beliefs (OECD, 2009, Creating Effective Teaching and Learning Environments: First Results from Talis, http://www.oecd.org/education/school/43023606.pdf.) Note: Constructs that are covered by the survey are highlighted in blue; single item measures are indicated by an asterisk (*)

**Box 5.13.** 'Out of field' teachers in Thailand

According to research undertaken by the Ministry of Education in Thailand, most foreign teachers are classified as 'out of field', meaning that they do not hold degrees in teacher training, nor did they major in their specialist subject. This means that, despite being able to communicate in English, TEFL teachers frequently do not have the appropriate subject knowledge or teaching skills. Many of the foreign teachers are tourists who generally teach only for short periods, thus creating the need for continuous recruitment. Teachers are often not conscientious about their responsibilities, behave inappropriately and resign from their positions without prior notice, despite their contractual agreements. As such, while hiring foreign teachers may help combat the shortage of English teachers in Thailand, this can also cause a number of inherent problems.

(Punthumasen, 2007)

likely that the answer to this question is no. With few formal entry requirements prescribed by TEFL agencies, no apparent regulation of the industry and chronic shortages of English language teachers in some parts of the world, it is clear that there will be instances when TEFL teachers are appointed inappropriately.

Borg (2003) emphasises that it is critical for teachers to be educated and well informed because prior to teaching many have inappropriate, unrealistic, or naïve understandings of teaching and learning. Furthermore, student teachers often have inadequate concepts of curriculum and programme design. There is a valid argument in questioning the value of a TEFL course that can be completed in a day or in 40 hours on a computer and it can be suggested that students who undertake courses of this type are likely to fall short of the pedagogic knowledge and experience required to effectively teach English as a foreign language.

Given that the roots of TEFL tourism lie in the concept of development, teacher skill shortages is a major concern. If teachers cannot meet the educational outcomes desired then the presence of TEFL tourism undermines the benevolent intentions which underpin it. It can be argued that placing a development agenda within a capitalist environment is asking for trouble. Profit-orientated organisations are unlikely to focus their efforts on the learning outcomes of the students, but rather they aim to make a rewardable financial return. As such, the calibre of TEFL tourists recruited becomes less of a concern, with 'bums on seats' being most important. This demonstrates the huge potential for the emergence of a number of negative impacts deriving from TEFL tourism, yet the issue has not been addressed in either the academic community or within the education industry. This notion of sustainability is further explored in Chapter Six.

U.A.E.
PAKISTAN
OMAN
Muscat
Karachi
Ahmedabad
Rajkot
Mumbai
Pune
Jaipur
Delhi
TIBET
C H I N A
Rutog
Chandigarh
Jaziral Masirah
Kuria Muria Is.
(Oman)
ARABIAN SEA
Abd al Kuri
Socotra
(Yemen)
Bargal
Aden to Colombo 2100
Panaji
Hyderabad
Bangalore
Chennai
BAY OF BENGAL
Yangon
Mumbai to Zanzibar/Dar es Salaam
SEYCHELLES
Victoria
Amirante Islands
Aldabra Is.
Diego Suarez
MADAGASCAR
Antananarivo
MOZAMBIQUE
Beira
Cargados Is.
(Mauritius)
MAURITIUS
Port Louis
Rodrigues
Saint Denis
REUNION
(France)
Capetown to Colombo 4449
Tropic of Capricorn
CHAGOS ARCHIPELAGO
SWAZILAND

# 6

# The Sustainability of TEFL Tourism

Niche tourism plays on the pejorative connotations that have accompanied the growth and development of mass tourism and its commonly associated negative impacts in relation to the environment, economy, political infrastructure and society (Novelli, 2005). For those seeking to utilise tourism as a mechanism for economic and social development, the niche tourism approach tends to provide a more sustainable and less damaging form. As a micro-niche tourism form, TEFL tourism therefore provides considerable scope for stakeholders to take full advantage of the potential positive impacts. It is important to note, however, that although TEFL tourism may not necessarily be associated with the frequently cited negative connotations of mass tourism, it is not exempt from negative impacts itself.

Described by the Brundtland Commission (United Nations, 1987) as 'development that meets the needs of the present without compromising the ability of future generations to meet their own needs', the concept of sustainability has increasingly become a staple term within the household environment and sustainable practices have indeed become everyday in many instances. This increased awareness is replicated within the tourism industry, where operators and individuals in general think far more consciously than they did in previous years. In 2005, the World Summit on Social Development identified three core areas that contribute to the philosophy and social science of sustainable development: economic development, social development and environmental protection.

While there is, of course, an element of responsibility within all three pillars of sustainability, it is the economic and social aspects that are the predominant concerns of this book, given that aside from the building of necessary infrastructure and everyday running of operations, TEFL tourism has little involvement with the environment. Despite the exclusion of environmental issues, the concept of sustainability within TEFL tourism remains a vast and complicated endeavour. Differences in geographical context along with economic, cultural and political ideologies add significant layers of complexity when attempting to assess the sustainability of the TEFL tourism industry. In fact, with so few studies focused on TEFL tourism, this is a very difficult task. This chapter provides an overview of sustainability considerations that should be taken into account by industry operators and academics. While lessons can be learned from tourism allies such as the volunteer tourism industry and from experiences in the wider economic field, to date there appears to be no specific documented evidence of sustainable (or otherwise) practices among the TEFL tourism community. Although this chapter enlightens readers in this area, there is an urgent need for further research to ensure the sustainable management of the TEFL tourism sector.

## TEFL as a Tool for Development

English is continuing to grow as the world's de facto lingua franca, and as an integral way to bring people together on a social, economic and political level. The ability to communicate in English opens doors for nations, communities and individuals, frequently providing opportunity for mobility and education. Economic growth is supported through international trade, and people from diverse backgrounds, to build a more positive intercultural dialogue. Governments from around the world are now recognising the importance of English in the development of their economies and societies and are seeking methods to make the learning and teaching of English in their education systems as effective as possible.

According to the British Council (2018b), new growth markets for English are largely outside the traditional countries of the developed world and they are growing at a rapid rate, predominantly as a result of globalisation and increased urbanisation. One of the strongest incentives for learning the language is the use to which it can be put immediately, socially, economically and culturally. The ability to communicate in English has considerable impact on economies overseas. English language skills frequently provide life-changing opportunities and promote prosperity and security around the world. In fact, it is argued that development efforts have now become inextricably linked, in governmental and academic circles as well as in the media, with English language education (Bruthiaux, 2002).

At an individual level, the relationship between English proficiency and gross national income per capita is a virtual cycle. While improving English skills generally results in an increase in salaries, which in turn gives governments and individuals more money to invest in English training, improved English skills also allow people to apply for jobs that they would not have previously been eligible for, thus raising standards of living. (United Nations, 2011). At a company-based or international level, English language communications facilitate business growth, as outlined in Chapter Four.

English language learning has traditionally held an important role in the educational design of many developing countries (Erling and Seargeant, 2012). Sometimes the pressure for the teaching and learning of English within a country has come from outside the country itself and at other times it is the result of the desire and aspiration for the acquisition of cultural capital. In the last two decades, however, there has been a shift in the reasons for developing English language provision, with a growing emphasis on English as a global language and its perceived role in providing access to both economic and social development. English language development can support a number of laudable initiatives such as ending poverty and hunger, providing basic education to all and improving child and maternal health, but these outcomes are extremely difficult to measure.

**Box 6.1.** English – changing lives for the better

Saranjeeve, a young boy living in rural Sri Lanka, was given the opportunity to learn English in primary school by a local NGO. When his parents could see he was making excellent progress with his English language learning they began to save as much money as they could to send him to university in the UK as they believed that this would provide him with the best employment prospects.

Unfortunately, they did not earn enough money so when Saranjeeve completed his secondary school education he found employment. While his English was not perfect, it was good enough for him to secure a job as a taxi driver for a tour operator. Every day he would pick tourists up from the airport and he would speak with them, learning more and more English each day. Having seen the potential for greater earnings, eventually he began to work for himself, providing custom-made tours and excursions for tourists.

His self-employment allowed him to make enough money to save enough to attend university and at age 20 he travelled to the UK to complete a foundation degree. Ten years later he has a successful tour operations business of his own back in Sri Lanka, he is teaching his own children how to speak English and he can afford to send them to a private school.

His success was entirely down to his ability to speak English, facilitated by the NGO which kindly donated time and resources to Saranjeeve's school when he was young.

**Fig. 6.1.** The Global Goals for Sustainable Development (United Nations, 2018)

One such initiative was the Millennium Development Goals (MDGs) which have since been followed by the 17 Sustainable Development Goals (SDGs). These are a set of shared aspirations and efforts to make the world a more equitable and sustainable place. While the SDGs are not legally binding, governments are expected to take ownership and establish national frameworks for the achievement of the 17 goals. Countries are required to follow up and review progress made in implementation of the goals. The SDGs recognise that ending poverty must go hand-in-hand with strategies that build economic growth, many of which also go together with English language capabilities. While the links between the goals and English language may not always be entirely transparent, their underlying importance is indisputable. For example, being able to communicate effectively can increase business and job prospects; thus this has the potential to raise people out of poverty (SDG1), which can reduce hunger (SDG2) and improve the economic situation in a given area (SDG8). A study by UNESCO (2012) identified the importance of language in achieving these goals. Language is at the centre of human activity, self-expression and identity and plays a key role in all aspects of life. Although this research promotes the use of local languages as opposed to using English, it does highlight the difficulties raised as a result of communication barriers and the impacts that this can have on development.

The role of the English language in sustainable development cannot be over-emphasised. Woolard (1998) stated that English has advanced as a valuable resource for international development. The associations being made are the perceived benefits to which the language can provide access because of its global status: 'Ideologies of language ... envision and enact ties of language to identity, to aesthetics, to morality and to epistemology' (p. 3). While the economic impacts of enhanced English language capabilities were made clear in Chapter Four, there are

**Box 6.2.** Language proficiency in Bangladesh

A project working to improve the situation of the Kol and Koda minority people in Bangladesh found that the communities were losing engagement with their mother tongue but were also not proficient in the dominant language, Bangla, nor were they proficient in English. This placed them at multiple disadvantages within education, business and general day-to-day life, where the indigenous population were struggling to communicate effectively in either language.

A joint project of SIL Bangladesh and Food for the Hungry placed importance on strengthening the mother tongue, while also focusing on enabling the communities to take ownership of the development process through participation. The project has engaged with a wide range of local ideas about language and language choices, while maintaining an emphasis on the importance of the mother tongue. To strengthen the local languages, the initiative has worked on developing local writing systems. It continues to work in the local languages to develop community institutions to bolster self-reliance, savings, skills development and advocacy for rights.

(UNESCO, 2012)

also social advantages. Being able to communicate with those from other parts of the world can be empowering for the host destination, enabling them to develop links which may in turn have positive economic impacts for the destination.

There are a number of studies which have examined the role of literacy in processes of poverty reduction, and its importance to the host community's capabilities and well-being (Street, 2001; UNESCO, 2005; Basu *et al.*, 2009). Further research has also explored the link between language use and development, demonstrating the way in which the languages spoken can influence economic status (Djité, 2008; Harbert *et al.*, 2009). While English certainly plays a role in poverty reduction, it is generally part of a larger picture and is not the answer in itself. Alongside the SDGs many governments, NGOs, charities, etc. will use the Sustainable Livelihoods Framework as guidance.

The Sustainable Livelihoods Framework was developed to help understand and analyse the livelihoods of the poor. While, like all frameworks, it is a simplification, it does provide a basis from which the full diversity and richness of livelihoods can be understood through qualitative and participatory analysis at a grass roots level. The intention of the framework is to be a versatile tool for use in planning and management from a top-down perspective. The aim is for the framework to act as an aid in helping to order complexity and make clear the many factors that affect livelihoods in particular situations.

English language development is intertwined within the fundamental aspects of human, social, natural, physical and financial capital, identified as 'livelihood assets' by the DFID.

The sustainable livelihoods approach is designed to empower poor people by viewing them not as victims, but as decision makers with their own sets of priorities. It employs a sector-by-sector view of development which is designed to accommodate the variety of economic activities an individual may rely on in order to subsist. The sustainable livelihoods approach, therefore, takes a holistic approach to poverty which considers a range of different resources permeating beyond income levels and productivity. The approach is, however, not without critique.

To begin, there is a lack of guidance regarding what makes one 'poor', thus making it difficult to determine which people the model aims to assist and which it does not. It can also be argued that the model does not provide enough emphasis on the informal structures and processes that affect access within the community. While local organisations may have the most noble intentions, they may not have the capacity to carry out the type of analytical research integral to the sustainable livelihoods approach to development. There is also the matter of working within differing cultural and political contexts, which must be taken into account in order to employ a practical framework. Lastly, while the framework does make note of gender considerations, attempts to increase the voice of women is difficult to successfully achieve.

There is now a consensus between both academics and linguists that the relationship between language and power is a mutual one.

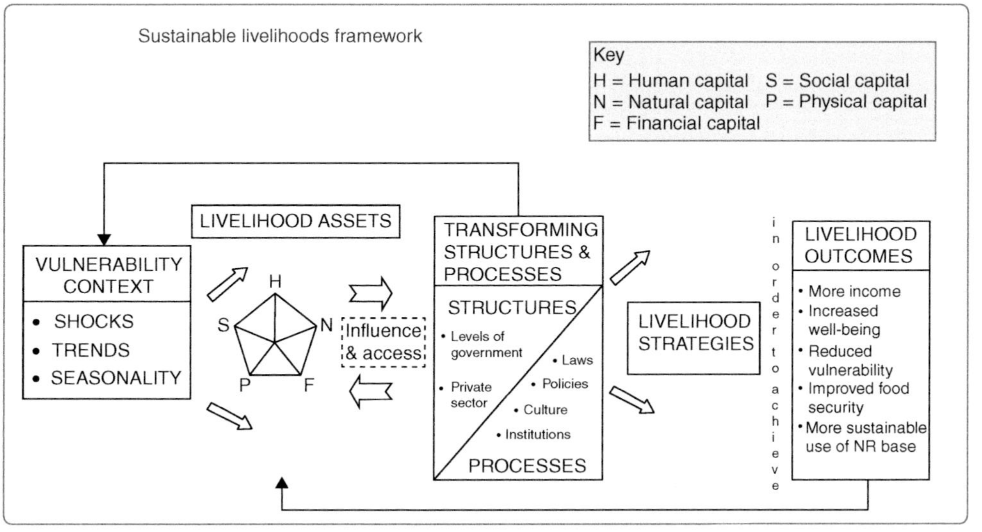


**Fig. 6.2.** The Sustainable Livelihoods Framework (DFID, 2000)

**Table 6.1.** An application of DFID Sustainable Livelihoods Framework to TEFL

| Livelihood asset | Explanation | Application to TEFL |
|---|---|---|
| Human capital | The amount and quality of knowledge and labour available in a household | Work opportunities may be enhanced through the ability to communicate in English |
| Social | Social resources, including networks for cooperation, mutual trust and support | Being able to communicate in English may facilitate the formation of relationships with English speakers |
| Natural | The quality and quantity of natural resources, ranging from fisheries to air quality | Money raised from TEFL tourism can aid in preservation of natural resources<br>Enhanced education can help enforce sustainable management of resources |
| Physical | The infrastructure, tools and equipment used for increasing productivity | Increased earnings produced through enhanced employment opportunities (see human capital) can help invest in infrastructure and public/private resources |
| Financial | Savings and regular inflows of money | Increased job prospects can enhance household income and be reinvested into the local economy |

The study of said relationship has shifted from proving its existence to probing and understanding its underpinnings and implications. Language can be considered as a communicative infrastructure, based on which the pillars of identity can be raised to contribute to the formation of an overall structure, which can be seen as power. In essence therefore, language provides the required identity by which a person or collection of persons can gain expected power. Language is an essential part of identity and allows for connections to be made with other members of various groups in society – ethnic, national, religious, gender, class, etc., which can have significant positive impacts on the development of a destination and its host community.

Language is delineated as a social practice (Fowler, 1975) which facilitates power relationships to be established and continued. Fairclough (1989) claims that power is not only built and sustained via force, but also via indirect

> **Box 6.3.** Escaping poverty in Bhutan
>
> Upadesh grew up in a rural village in Bhutan. His parents worked hard on their farm to save money in order to provide him with the best education possible. He did very well at school and was fortunate enough to be able to travel to the UK to undertake a degree. Upadesh returned to his home country and managed a small hotel and tourism business, eventually becoming a successful entrepreneur.
>
> Unfortunately, the relationship between him and his family weakened, and although he continued to provide financial support for them, he began to view himself as superior, largely due to the fact that he had considerably more financial capital than they did, owing largely to his ability to communicate in English and to have completed his education in the UK.

ways such as the use of language. There are many examples where language is used as a tool for maintaining status and power, for example between the educated and uneducated, high fliers working for multinational corporations versus those working in local communities on low salaries or between family members. TEFL can be a route to achieving this 'power', but at the same time can also widen the gap between the powerful and the powerless.

One way to comprehend the relationship between language and power is by examining the work of Foucault. The French postmodernist has been hugely influential in helping people to understand the concept of power, leading away from the ideas that actors predominantly use power as an instrument of coercion, and the discrete structures in which said actors operate, towards the idea that 'power is everywhere' (Foucault, 1991; Rabinow, 1991). According to Foucault, power is what makes each person the way that they are; power is diffuse as opposed to concentrated, embodied and enacted rather than possessed, discursive rather than purely coercive, and constitutes agents rather than being deployed by them (Gaventa, 2003). Foucault describes power as a type of 'metapower' or 'regime of truth' which pervades society and is in constant flux and negotiation. The term 'power/knowledge' is used to signify that power is constituted through accepted forms of knowledge, scientific understanding and truth.

Foucault (1991) believed that truth is produced only by virtue of multiple forms of constraint, which in turn induce regular effects of power. Both in the TEFL domain and outside, different societies are subject to their own regimes and politics of truth. Said 'general politics' and 'regimes of truth' are evident and constantly reinforced through the local education system, the media and the relevant political and economic ideologies. This adds an additional level of complexity to the weighty debate over the relationship between language and power, indicating that this relationship may pose different states in different cultural and political regions.

It is recognised that power is not just a negative, coercive or repressive force that pushes people to do things against their wishes, but can also be a necessary, productive and positive force in society (Foucault, 1991; Gaventa, 2003). The evidence for this with regard to English language capabilities is indisputable, with increased job prospects, enhanced local and global economics and improved livelihoods being common outcomes resulting from improvements in English language communications in marginalised or less developed nations. While the benefits have been covered in this book so far, however, less has been discussed of the negative impacts.

It is evident that language has the ability to construct and manipulate concepts of power in society (Fowler, 1975; Kramarae *et al.*, 1985). The way in which power is constructed by institutions and individuals can encourage globalisation, which can change societal practices and everyday life in indigenous communities, affecting the authenticity of the destination and its peoples and encouraging increased notions of 'sameness' (Mayr, 2008). Research in the area of globalisation associated with TEFL is most certainly in its infancy. While there are some studies beginning to emerge examining pedagogical globalisation or the role that English language plays in an ever-increasing world of interconnectedness, there is little explicit evidence demonstrating the role that the practice of TEFL might have in diluting local cultures and traditions.

Nonetheless, there are countless examples of the cultural impacts which commonly occur when a westerner is placed in an authentic, indigenous community. Examples include the loss/limited use of traditional languages and the

**Box 6.4.** Authenticity and TEFL

Turner (1994) stated that the very existence of tourism rules out the possibility of an authentic experience. While there are many debates about whether authenticity is or is not what the tourist seeks (Boorstin, 1964; Cohen, 1972; MacCannell, 1973), there is strong evidence to suggest that by placing a tourist into an otherwise 'untouched' community, globalisation will begin to occur and thus authenticity will begin to cease.

In the TEFL environment, therefore, it is inevitable that the introduction of a western TEFL teacher who is partaking in TEFL tourism will lead to cultural changes, whether intentional or otherwise.

While this might sound like a negative impact, it raises a fundamental question with regard to how authenticity is actually and operationally defined. Cultural changes are an inevitable part of cross-cultural relations, but does this mean that communities should be isolated from the rest of the world to allow for them to 'maintain their authenticity'? One could argue that in actual fact it is *inauthentic* to purposefully isolate communities in this way and that globalisation is an inevitable and natural part of societal evolution.

Either way, there is a responsibility for the TEFL industry to mindfully manage cross-cultural interactions and there is a need for further investigative research in this field.

'copying' of western practices – i.e. using particular turns of phrase, dressing in a particular way or behaving in a certain manner. Such impacts can filter through to all aspects of society, resulting in changes in values, operational practices, ideologies and relationships.

In terms of development, it can be seen that there is huge potential within the TEFL tourism industry. The previous pages have only touched on what is an important, yet extremely complex, area. What is most surprising is that there is very little literature presented in this field with specific regard to TEFL, despite the primary intention for most nations' implementation of TEFL tourism being as a means to development. This is somewhat worrisome, given the time, infrastructure and money that is often invested in TEFL, compared to the consideration that appears to be taken of some of the wider-reaching impacts, particularly those of a negative nature. This is explored further later in this chapter.

## TEFL Tourism as a tool for Poverty Reduction

Pro-poor tourism (PPT) is not a sector of tourism per se, but rather it is an approach aimed at increasing the net benefits for the poor. PPT strategies are concerned specifically with impacts on poor people, although it is important to recognise that others are likely to benefit also. The focus is not to expand the overall size of tourism, but to unlock opportunities for specific groups within it (Overseas Development Institute, 2001). Simply put, the intention of PPT is to direct profits back into the community through aspects such as local employment, leasing of local land, using local food and other resources, operating local businesses or, in the case of TEFL, enhancing career prospects.

It is important to note that proponents of PPT are not anti-capitalist. Rather, strategies are designed to incorporate the poor into capitalist markets by increasing the employment and entrepreneurial opportunities available to them. In order to achieve this, it is crucial that PPT is integrated into the wider tourism system and successful PPT depends on the poor having access to markets, on the commercial viability of PPT projects, on a policy framework that provides a secure investment climate (including access to land), and on effective stakeholder cooperation and strategy implementation. In the context of TEFL tourism, this means that it is imperative that public–private relationships and partnerships are developed and maintained from a top-down perspective as well as at grass roots level, between educational institutions, TEFL tourism providers and local governments.

Similar to fair trade, PPT is essentially a form of market intervention. Rather than being concerned with the absolute (net) benefits received by the poor, it focuses on the concept of development. PPT strategies are considered to include aspects such as infrastructure gains, capacity-building and training, and empowerment. While the main focus is often on direct employment, TEFL tourism works slightly differently in that it prepares and equips the individual for employment, rather than providing it. Furthermore, it encourages micro-enterprise development and livelihood diversification, which can have a substantial effect on the community as a whole given the potential for tertiary benefits. The ODI (2007) has divided PPT strategies into three prominent categories encompassing developing pro-poor policies, partnerships/processes and increasing economic benefits for the poor, and tackling non-economic impacts. The issues associated with these categories are, however, most certainly not linear as poverty is a multi-dimensional phenomenon. Therefore, in order for strategies to be effective, stakeholders must address not only poverty alleviation, but seek to eliminate the underlying causes.

In assessing the appropriateness of utilising TEFL tourism as a means of PPT, there are two fundamental underlying theories to consider: modernisation and dependency. In the context of modernisation theory, pro-poor TEFL tourism can be viewed as an effective method of reducing poverty by generating foreign exchange, increasing a less economically developed country's (LEDC) government revenues through tourist expenditure, attracting foreign development capital, enhancing local employment prospects and promoting 'modern' western values and practices.

Modernisation theory holds that LEDCs develop as they adopt western 'modern' approaches, an arguably inevitable part of globalisation as discussed in the previous section. Researchers inspired by Durkheim's functionalism theory argue that a society will autonomously eliminate impractical elements during the process of development (Tipps, 1973) and that traditional values and practices may become deemed dysfunctional or old-fashioned, and will thus be substituted with wider and consistent social, cultural and political changes (Rostow, 1960; Hoogvelt, 2001).

Modernisation theory ascribes economic growth to socioeconomic development and the ODI (2007) quantified the indirect positive impacts of PPT in local economies by about 50% to 90%. This builds on the fact that there is no disputing the potential economic benefits of tourism. The ILO (2012), for instance, reported that international tourism is the third export sector in LEDCs, accounting for about a third of all exports. Furthermore, the tourism industry proportion of exports in services accounts for 29% worldwide and 52% for LEDCs (World Tourism Organisation and European Commission, 2013). While these figures appear impressive, there is little indication prescribing to which tourism sector revenues are attributed, the extent to which economic leakage may occur and at which costs these benefits occur.

In order to capitalise on said economic impacts, it is imperative that scholars and industry practitioners are able to identify and examine these on a more specific level. This is one reason for the publication of this book – to facilitate a grounding point from which TEFL stakeholders can underpin their thought processes, decision-making and strategy implementation. While there is undoubtedly a direct correlation between tourism and the creation of economic net benefits for the poor, the extent to which this applies to TEFL tourism is unclear. Tourist spending, for example, will almost certainly be lower for a person who is living in and is integrated in the local community – through eating local food for instance and from having limited free time for leisure pursuits – than it would be for a fully-paying tourist in holiday mode. The multiplier effect, on the other hand, may well be superior for communities which host TEFL tourism in comparison to the average tourism destination, as a result of the focus on skills development of locals which aims to enhance employment prospects. Consequently, if a greater number of the community are in meaningful employment then they are likely to reinvest money earned into the community through personal spending, donations in kind, taxes paid and so forth.

The concept of development through tourism was further explored in 1993 by Oppermann through his research based in Malaysia. He categorised tourists as being either 'active' (those who stayed in more than four places) or 'passive' (those who visited a maximum of four locations). He found that 'active' travellers contributed substantially to regional development while 'passive' tourists tended to strengthen existing spatial disparities. TEFL tourists who book their experiences through a western TEFL agency, stay in multinational hotel chains or eat

**Box 6.5.** Western-owned TEFL tourism operators – what is the benefit?

There are now a number of TEFL agencies which act as a form of tour operator, packaging together TEFL elements for the tourist. This typically includes a TEFL training programme/qualification, accommodation during training, in-country transfers, pastoral support and guidance and sometimes may include flights and a job placement.

While there is a strong argument to suggest that local people or experienced TEFL teachers would be best suited to organise and put together TEFL packages of this nature, this opportunity has been capitalised on by a number of major tour operators with little or no previous experience in this domain.

Often operating under subsidiary organisations or integrated companies, the major tour operators who previously focused their operations on the mass market are now reaping the benefits of TEFL tourism. Unfortunately, these rewards tend to end up in the hands of western CEOs and their staff as opposed to the community which often desperately needs them.

at foreign-owned restaurants so are unlikely to contribute to poverty alleviation, but instead strengthen existing inequalities in LEDC destinations. This suggests an interesting ethical conundrum regarding the PPT phenomenon and TEFL. Indeed, a number of TEFL tourists may travel to LEDCs with the hope of contributing to poverty reduction through their participation in education provision, yet they are undermining these benefits by contributing to tourism expenditure leakage through the choices that they make outside of the classroom. Of course, the best solution to promote PPT is to support the establishment of locally-owned touristic enterprises. However, many international tourists demand 'luxury' services such as air conditioning or western consumer goods which local businesses may not be able to afford. As a result, western-owned resorts possess a substantial competitive advantage over local tourism enterprises which seriously undermines the effectiveness of PPT initiatives.

Dependency theorists typically view international tourism as a form of neo-colonialism used by western nations to perpetuate a pattern of dependency over LEDCs (Oppermann, 1993). This is especially prevalent in TEFL tourism given power relationships imposed as a result of the student–teacher dichotomy, supported by the fact that TEFL teachers are most likely to be western individuals as a result of the inherent English language capabilities. As a result of such dependency, Oppermann suggests that international tourism reinforces existing socioeconomic regional disparities within LEDCs through its enclavic structure and its orientation along traditional structures. Essentially, dependency theory ascribes the development of western countries to the exploitation of LEDCs, seen as peripheral to the core nations based in the West (Hoogvelt, 1997). The flow of said tourists from the so-called core to peripheral countries and the management of the tourism industry is subject to various control mechanisms imposed by western institutions (Britton, 1989), resulting in a dependency being created between LEDCs and western nations. While for many tourist forms, such dependencies may not be evidently visible to the untrained eye, they are quite apparent within TEFL tourism. The TEFL sector is largely reliant upon western teachers to operate in the classroom and on western agencies to organise teacher placements, recruitment, training, etc. It is indeed a reality that in the majority of instances PPT, enterprises including TEFL tourism substantially depend upon foreign investment. Arguably, as western businesses are driven by profits, the benefits of PPT are questionable. This is addressed in more depth later in the chapter.

The last theoretical perspective worth noting is that of structuralist theory. The intention of structuralist theory in social anthropological examination, such as of tourism, allows for meaning to be produced and reproduced within a culture through various practices, phenomena and activities that serve as systems of signification. Grounded on dependency theory principles, structuralist theory in the context of tourism ascertains that current tourist destinations in LEDCs serve as 'pleasure peripheries' for core western countries (Balaam and Dillman, 2013). This presupposes that PPT is perceived as a mere tool for furthering western interests in its 'pleasure peripheries' on offer. There is some evidence of this within TEFL tourism, where tourists are motivated and indeed incentivised by TEFL agencies to take part in TEFL tourism, not as a result of philanthropic or career-based motivations, but instead as a means to enjoy the leisure facilities available to them in the TEFL destination. This is discussed further in Chapter Seven.

In essence, while there are connections between TEFL tourism and PPT, there is evidence that this may not always yield the positive outcomes desired. This is, however, just one perspective on the negative impacts which may result from TEFL tourism and while there is little documented regarding the perceived or actual negative outcomes to date, it is important that such concerns are raised in order to allow for mitigation and appropriate management.

## The Impacts of TEFL Tourism

So far, this book has hailed the benefits of TEFL tourism in terms of employment generation and development prospects, but it has also been critical of aspects such as the difficulty in standardisation, regulation and monitoring of an industry of such a scale, teacher skills and suitability and dependency. It is difficult to say exactly what negative impacts there may be, given the sparsity of research on TEFL as a form of micro-niche tourism. However, we can draw on existing research addressing the likes of the mass sector and macro-niches that share similarities with TEFL tourism such as volunteer tourism. Inherently, the tourism industry is reactive as opposed to proactive, putting effort in place to sustainably manage different aspects of the industry once an issue has been observed. Given that the tourism industry has grown to such an extent, there is an opportunity for TEFL tourism stakeholders not only to learn from their own mistakes, but also from those of others.

There is an array of documented evidence surrounding the negatives resulting from tourism, typically categorised as either environmental, economic or social. The positive economic impacts have been discussed in many parts of this book and are fundamental motivations for both public and private sector involvement. There is also an inevitable social involvement which needs to be taken into consideration. Generally, optimist views towards TEFL tourism have been complemented by a fairly uncritical approach towards the sector, when in reality it should be critically analysed just like any other form of tourism. Although TEFL tourists' benevolence may appear advantageous, there are likely to be numerous possible negative impacts that currently receive little attention. The following section, therefore, is based on a review and analysis of the relevant tourism literature and highlights a number of potential negative impacts of TEFL tourism, questioning the somewhat idealistic depiction of the sector. The intention is to inspire future critical research into TEFL tourism and to serve as a guiding framework.

Fig. 6.3 shows the anticipated advantages and disadvantages of TEFL tourism to the host community. As has been reiterated throughout this book, there is a dearth of available literature examining the TEFL industry as a form of tourism and thus Fig. 6.3 acts as advisory content to stakeholders of TEFL tourism as opposed to being explicit. The following sections simply provide insight into the anticipated impacts associated with TEFL tourism.

### Social impacts

Understanding the social impacts of TEFL tourism on communities is extremely important for stakeholders at all levels so that action can be taken to reduce the likelihood of a community backlash against TEFL tourists and associated tourism development. Given that the TEFL learners and surrounding community are a fundamental part of the TEFL tourism 'product', societal attitudes and behaviour have a sizable impact on the success or otherwise of TEFL tourism in the destination. From a broad perspective, research on the social impacts of tourism on communities is substantial and ongoing and while advances have been made in the area, the consensus is that research has not addressed some of the deep-seated issues faced by tourist destinations, particularly for niche tourism forms such as TEFL tourism.

Tourism has long been considered a means from which hosts and visitors can learn from each other, through direct and indirect interaction.

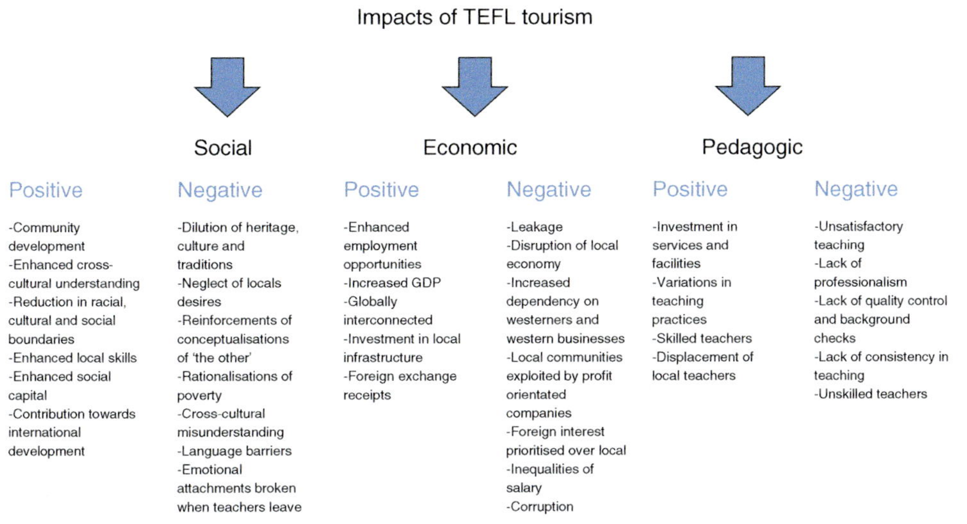


**Fig. 6.3.** Impacts of TEFL tourism

This is especially true for TEFL tourism, where educational outcomes are not only present in an informal sense from learning through travel and interactions generally, but from organised and intentional educational provision. Interactions between locals and tourists allows for the emergence of new ideas, values and motivations for social and economic progress (Brown, 1998; Bersales, 2005; Nyaupane *et al.*, 2006). Tourism can revitalise a community's cultural life as art and traditions are an attraction for foreign visitors, but it can also have detrimental effects.

Many practitioners and academics utilise the framework developed by Doxey in 1975 to assess the evolution of tourism with regard to the host community and their views and involvement of tourism development in the area. Doxey's Irritation Index, also known as 'irridex', examines the evolution of the relationship between tourists and the locals. It posits that communities pass through a sequence of reactions as the impacts of an evolving tourism industry in their area become more pronounced and their perceptions change with experience. He suggested that the attitude of members of the host community evolves through four distinct stages, which are in parallel to the phases in the evolution of tourism at a destination in Butler's (1980) tourist area lifecycle (exploration, involvement, development, consolidation, stagnation and decline or rejuvenation). These stages are implicitly accompanied by increasingly adverse social impacts as the nature of tourism in the area develops (Fig. 6.4).

Due to the recent emergence of TEFL as a form of tourism, it is likely that many TEFL tourism destinations are currently experiencing Doxey's first stage of societal emotion: euphoria. It is during this stage that people are generally enthusiastic and thrilled by tourism development. They welcome the stranger into the classroom and community and there is a mutual feeling of satisfaction. There are opportunities for locals, money flows in along with the tourists, and the level of English language capabilities in the community increases.

This is then followed by the apathetic stage, of which there is a growing body of evidence in certain popular TEFL regions such as South-East Asia and some African countries. As the TEFL tourism industry expands, members of the community begin to take the tourist for granted. Educational institutions have probably worked with a number of tourists by this stage which brings inconsistencies in teaching practices and student–teacher relationships. The tourist rapidly becomes a target for profit-taking, and there is an emergence of TEFL agencies, who are often western-owned, seeking to capitalise on opportunities that arise from this new tourism marketplace.

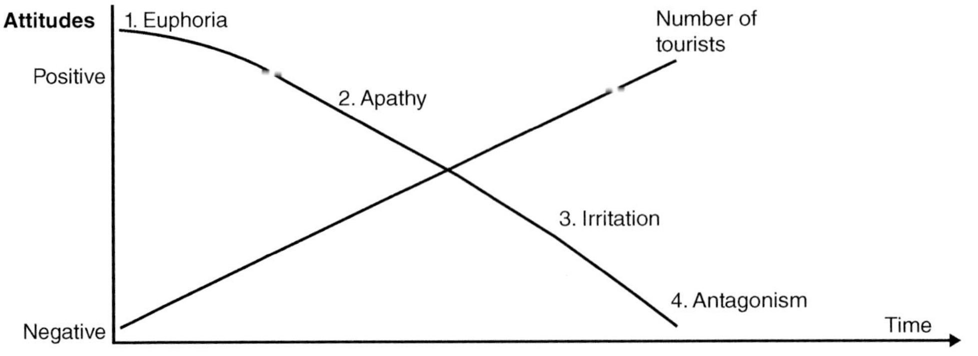


**Fig. 6.4.** Irritation Index (Doxey, 1975)

The third stage of Doxey's (1975) model suggests a stage of irritation. In the context of TEFL tourism, it is likely that at this time educational institutions are beginning to realise that the never-ending cycle of new TEFL tourists working in their organisation has its drawbacks – lack of consistency, failure to develop strong ties with the learners and to identify with their particular needs, being unaware of what has already been taught by previous teachers and so on. Student progress does not continue to improve at the rate that it did at the beginning. Recruitment of TEFL teachers is no longer an issue at this stage as there are now several TEFL agencies operating in the area, although the financial benefits are not always redirected into the community, but are retained by the western-owned operators.

This then leads to a stage of antagonism. Seeking more business opportunities and greater profits, TEFL agencies recruit and send tourists who may not have the required skills, personal qualities or cultural empathy to satisfactorily operate and integrate with the host community. Tourist enclaves start to form where they gather together in western-operated restaurants and bars, leaving little economic benefit for the local people. The price of accommodation rises as tourists begin to occupy apartments and houses, and taxes increase to fund areas such as airport development. Furthermore, cultural erosion starts to play a significant role and children begin to watch western TV shows, read western books or use western phrases that were previously unknown to them, as a result of learning in their English classes. The host community begins to feel aggrieved and antagonised by the TEFL tourists.

While Doxey's model may appear somewhat simplistic, it does correlate with many examples of social and emotional responses to tourism across the world. It therefore is appropriate to apply this model in the context of TEFL tourism in an attempt to warn TEFL stakeholders of the potential implications of TEFL tourism development in a destination.

TEFL tourism can also impact social capital. The concept of social capital became fashionable only relatively recently, but the term has been in use for almost a century while the ideas behind it go back further still. Social capital is defined by the OECD as 'networks together with shared norms, values and understandings that facilitate co-operation within or among groups'. The measurement of social capital in TEFL tourism destinations has the potential to provide valuable insights into the social networks and links that individuals and communities have, and importantly how these networks and links can be utilised to contribute to positive outcomes. Social capital may lead to enhanced social and economic well-being. Alternatively, it is just as likely that economic and social well-being lead to enhanced social capital, demonstrating a causal effect which runs in both directions (Putnam, 2000). Social capital originates and operates from a variety of different sources within a community. It is built in families, schools and other educational institutions, businesses, civic institutions as well as in the local community. All of these sources of social capital are important.

Although social capital is largely associated with positive outcomes, it is important to recognise that there are potentially negative outcomes too. The OECD (2001) notes that particular forms of bonding social capital have the potential to impede social cohesion in certain circumstances, an area that is particularly relevant to the concept of TEFL tourism given the inherent bond that is developed between the teacher and the student. The constant exposure to new teachers who stay for only a short period of time is likely to have adverse effects on children's ability to develop and maintain emotional bonds with individuals and could have greater impacts on their psychological well-being (Box 6.6).

A further way in which social capital may exert a potentially negative influence is that in some communities it may impose conformity and social division. Given the potential for new doors to be opened for those who develop their English language speaking capabilities through TEFL tourism, particularly in terms of employment opportunities and international relations, this could create societal divisions. Increased earning potential and contact with other cultures and communities may affect the behaviour and attitudes of individuals, with consequences for the make-up of the society and its 'authenticity'. This once again brings the concept of globalisation back to the forefront of this discussion. Globalisation is accused of destroying the sociocultural identity of local communities and indigenous values, traditions and lifestyles (Reisinger, 2009). Areas visited by tourists in developing countries become no longer authentic, but rather disinfected and 'McDonaldized' (Paramo La Haine, 2008). Diversity gives way to efficiency and local culture to global culture. A small village, which is facing many economic problems, can turn itself round by capitalising on the new skills acquired by those involved with TEFL tourism.

One final and important note on social capital is that there are links between the level of educational attainment in a society and its economic development. Participation in education is commonly the basis for building positive values that characterise social capital such as trust, acceptance, reciprocity and cooperation. Effective education can also assist in reducing problems such as unemployment, poor health and crime. Education, both in the TEFL domain and outside of it, may therefore be considered both a potential element of, and an outcome of, social capital.

## Economic impacts

There are inherent links between social capital and economic inputs and outputs and it can be argued that negative economic impacts can be the cause of subsequent negative effects on social capital. Meanwhile, economic improvements as a result of TEFL tourism can also create social and cultural benefits. In fact, it is often the economic impacts that are the main motivating factor for governments, companies and individuals to be involved with TEFL tourism.

Travel and tourism is an important economic activity in most countries around the world. There are a variety of tourism forms, of which the exact economic impacts are not necessarily as clear, but it is evident that, from a broad perspective, the benefits are indisputable. Economic benefits consist of those that result directly from tourism activity, such as hotel

**Box 6.6.** Emotional attachments and broken bonds in orphanage tourism

Orphanage tourism has been subject to criticism in the media in recent years and there is growing awareness of some of the negative impacts that it can have on a community and on the individual children living in the orphanage.

One such negative impact is that of broken bonds.

Children housed in orphanages often lack the loving and caring environment that they would have in a happy home. It is therefore no surprise that they 'cling on' to volunteers who work with them in the orphanages.

The problem arises, however, when the volunteer leaves, breaking the emotional bond that was established with the child.

Volunteer placements can start from as little as a week, meaning that children could be exposed to a great number of people throughout a given year. There is evidence that this is psychologically damaging for the child and that it can have lifelong consequences on their social abilities and their psychological health.

revenue or entrance charges to a museum, and from indirect tourist receipts through activities in support of the tourism industry, such as taxi drivers, fishermen who serve the tourist restaurants or construction companies used to build hotels. There is the potential then for said economic benefits to be multiplied, whereby stakeholders spend monies earned within the local economy and thus these monies can be reinvested in aspects such as social services, education or healthcare. Benefits of this type are found throughout the tourism industry, but the TEFL tourism sector is unique in that it not only provides the potential for economic benefits in the immediacy, but also in the long term, given that the intention is to enhance the employment prospects of the local community, which in turn boosts the local and national economy. TEFL tourism, therefore, provides a unique opportunity for the economic benefits of tourism to be capitalised and built upon, enabling maximum benefit to the community (Fig. 6.5).

Unfortunately, as with any capitalist venture, priorities are not always placed with the community, but instead with the profit-orientated TEFL stakeholders such as the agencies who commodify the TEFL product, putting it together to resemble a type of holiday 'package'. There is a distinct paradox between the capitalist marketplace and the will to 'do good'. The prolific nature of the industry is demonstrated, for example, through the company i-to-i which offers TEFL placements as part of its provision and, despite its individual branding, is owned by the profit-maximising organisation First Choice Holidays which is typically associated with modern package holidays. It is, therefore, no surprise that TEFL teaching opportunities provided by such agents have strong associations with the traditional package tourism model, as discussed in Chapter Four.

The monetary exchange for doing good poses several philosophical and ethical questions and it is argued that monetary gain is not appropriate in a world of benevolent intentions (Tomazos and Cooper, 2012). Some researchers have suggested that organisations are tapping into demand and are actively exploiting niches, such as TEFL teaching, by charging high prices (Tomazos and Butler, 2009; Keese, 2011), with questionable benefits to the hosts (Benson and

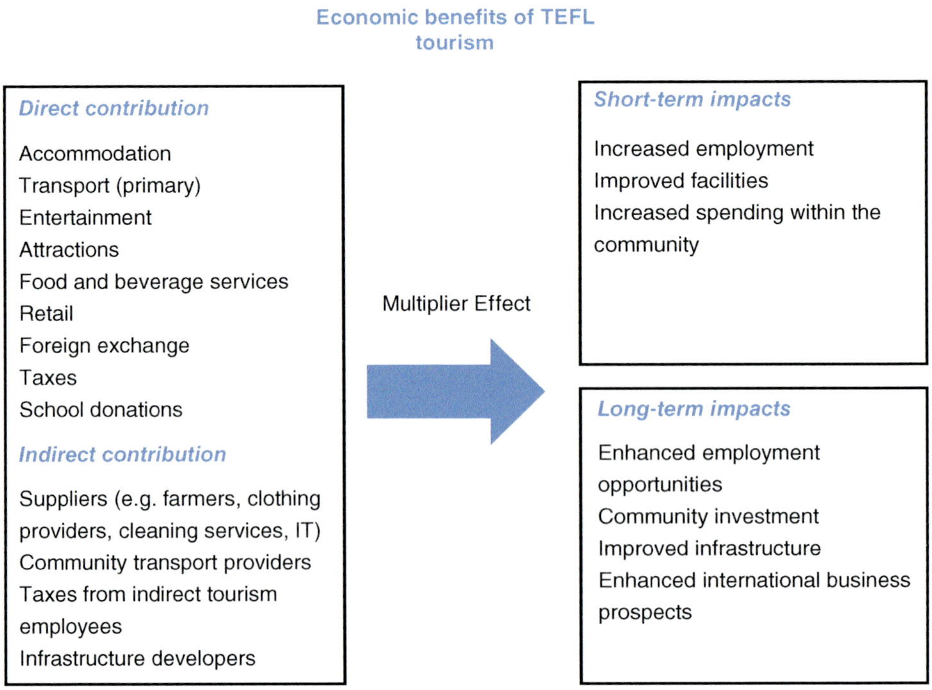


**Fig. 6.5.** Economic benefits of TEFL tourism

Wearing, 2012). Although it is not the intention of this book to analyse the commercial enterprises involved with the TEFL industry, it can be argued that TEFL stakeholders, most importantly the TEFL teachers, have a moral right to understand the ethos and intentions of the organisation to which they are choosing to commit. To date, there is no clear distinction between the types of agents offering TEFL placements and the differences in their services and values. This is akin to volunteer tourism, where it is argued that the ever-evolving marketplace contributes to an ambiguous industry. With such variations in terms of size, ethos and business, there are concerns regarding the value of projects promoted by these agents (Tomazos and Butler, 2009). Some agents have a broader knowledge and understanding of tourism and development (and by extension TEFL teaching) than others (Wearing and McGehee, 2013). However, the current lack of regulation of the industry leaves the door open for opportunists (Tomazos and Butler, 2009).

The ambiguous nature of TEFL organisations, coupled with the lack of regulation of the wider TEFL industry, gives rise to the question of whether those organisations with predominantly financial motivations will recruit the best teachers to help a country to improve its social and economic development. Chapter Four demonstrated the way in which many organisations provide a TEFL or teacher training course as part of the package. However, there are no details of the awarding bodies or levels of qualification acquired and some organisations do not specify whether training takes place online or face-to-face. This further supports the claims made regarding the ambiguity of qualifications held by TEFL teachers. Akin to this are the essential criteria for TEFL applicants. As with the volunteer tourism industry, many packages have minimal or no requirements regarding the skillset needed to participate (Brown and Morrison, 2003; Guttentag, 2009), a factor which can subsequently negatively impact on the quality of work undertaken (Guttentag, 2009; Benson and Wearing, 2012) and undermine the positive impacts desired for the host.

Unfortunately, in the past, tour operators have sometimes neglected their sustainability responsibilities, arguing that they are simply intermediaries between customers and service providers and that destination impacts are the responsibility of the subcontracted suppliers or the local authorities. While stakeholders clearly share the responsibility, most tour operators now understand that it is their obligation too as they work so closely with both tourists and tourism service suppliers (Swarbrooke, 1999; Budeanu, 2005). In addition, several researchers indicate that tour operators are in fact the key to achieving more sustainable forms of tourism (Swarbrooke, 1999; Wijk and Persoon, 2006; Frey and George, 2010).

This is a positive re-evaluation of the roles of tour operators; today, a large number of companies take a proactive approach and have started to develop sustainability plans. It is difficult to say to what extent tour operators offering TEFL products adopt such an approach, although given that it is common for large organisations to operate subsidiary organisations which focus on TEFL tourism, it is reasonable to assume that this area of the business will be encompassed within the bigger picture.

In order to encourage tour operators worldwide to make a corporate commitment to sustainable development and to make considerations of the environmental, social and cultural impacts an integral part of the design and operation of their tours and the conduct of their business activities, the Tour Operator Initiative (TOI) was developed. Supported by the United Nations Environment Programme (UNEP), UNESCO and the World Tourism Organisation in the year 2000, the initiative was developed to encourage tour operators to commit themselves to the concept of sustainable development as the core of their business activity by working together through common activities to promote and disseminate methods and practices compatible with sustainable development (TOI, 2005).

While there may be little environmental involvement in the practice of TEFL (outside of the use of transport and energy used), there are several areas warranting consideration by TEFL agencies and stakeholders to ensure sustainable management. According to Carbone (2004) and TOI (2007), key operating areas include: internal management; product development; supply chain management; customer relations; and cooperation with destination. Specific to TEFL, areas worthy of consideration may include (but are not limited to): recruitment of local versus foreign

**Box 6.7.** Tui Sustainability Strategy

In attempt to demonstrate their proactive approach to sustainability, Tui have an ambitious 2020 sustainability strategy entitled 'Better Holidays, Better World'. It is built around four core pillars to help shape the future of sustainable tourism, demonstrated below.

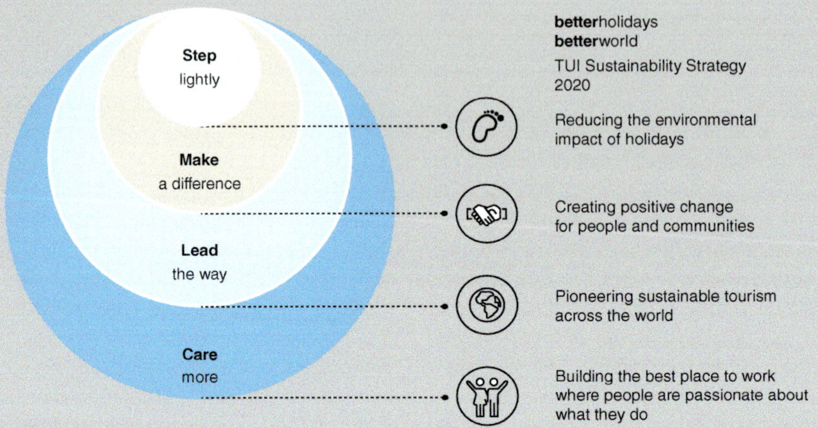


(TUI, 2018)

staff; whether TEFL tourist salaries are in line with local salaries; whether products and services required are sourced locally; how much money is retained by the destination/community; and whether the product offered is suitable for the TEFL tourist and the community.

While it is difficult to determine precisely how organisations, whether involved with TEFL or otherwise, distribute the money that they make from tourists, it does demonstrate good corporate social responsibility if they provide some indication of this, as Projects Abroad does. But unfortunately, distribution of capital is not the only negative economic impact resulting from TEFL tourism. Another area of concern lies with the very notion of sending tourists to act as TEFL teachers. With roots in western colonialism, the white tourist remains frequently viewed as superior, whether their English language is better than that of a native TEFL teacher or not. As a result, TEFL tourists are often paid more than their local counterparts or, at the other end of the spectrum, they volunteer their time at the expense of the local staff member who is subsequently made redundant.

The issue here is the impact on the local community. To begin with, the economic benefits that would have existed prior to the introduction of TEFL tourism (i.e. the local teacher salary) have ceased, but the negative implications permeate beyond this, with resentment and unhappiness from within the host community. Wage disparities are often a taboo topic, especially when power relations are involved, but this is an important element that needs considering. Indeed, the tourist may remain completely unaware that their position is the direct result of a local teacher becoming unemployed, but such naïvety is no excuse for the TEFL agency and other associated stakeholders. Agencies have a moral and ethical responsibility to ensure that by deploying western teachers they are not encouraging negative impacts such as this, whether intended or otherwise.

The displacement of local teachers is a genuine concern in the TEFL tourism sector. This refers to the forced movement of local teachers from their occupational activities as a result of the arrival of foreign teachers. What's more, there is a distinct possibility of gentrification as a result of TEFL tourism, where the wealthier western tourists move into residential areas previously solely occupied by the local community. Tourists may take up valuable accommodation

**Box 6.8.** Money spent on TEFL tourism – where does it go?

A recent article in the *Guardian* (2016) criticised TEFL agency Projects Abroad, stating that the organisation charges a minimum £1,145 to cover the costs (not including flights or visa costs) of two weeks' teaching in Cambodia, which is enough to pay a local teacher for more than a year. This demonstrates that there can be a lot of money to be made from such entrepreneurial ventures and in many cases the consumer does not know how their money is paid out. Projects Abroad has tried to clear this matter up by providing further details of how the money is distributed (see graphic below), although this is still somewhat vague. Projects Abroad claims that at least half of the fee from each volunteer is spent in destination directly, and that the money is used for the time they allocate on a project, as well as for the long-term planning and monitoring of that project.

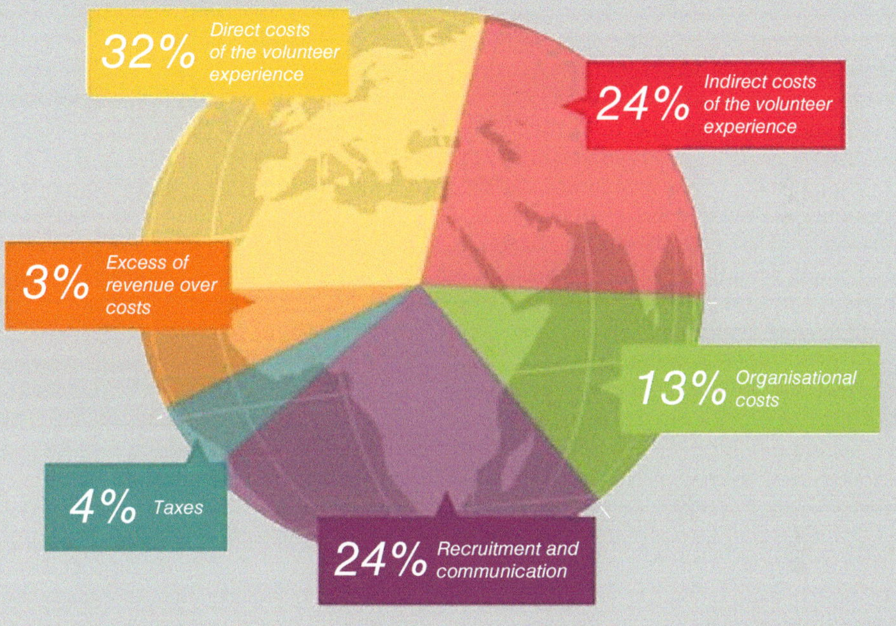


(Projects Abroad, 2018)

**Box 6.9.** Foreign vs local wage disparities

A survey of 1,300 local and expatriate workers found a wage gap that ranges from 400–900% and causes significant resentment among local workers. Local employees frequently have comparable skills and qualifications, undertaking the same job. In some instances, they have a better knowledge or a more responsible position than their higher-paid colleagues.

However, it is a common occurrence that Westerners are paid more than those from developing countries, despite working for the same company, in the same place, doing the same job.

An [Economic and Social Reaearch Council] ESRC-funded project called Addup (Are Development Discrepancies Undermining Performance?) studied the effects of the wage gap between local and international staff working in lower-income countries. The results demonstrated that disparities created significant feelings of workplace injustice which can damage relationships in the workplace and thereby interfere with employment effectiveness.

(*Guardian* online, 2018b)

and, because of the laws of supply and demand, may cause rises in rental or sale costs, subsequently marginalising the locals who can no longer afford to live in the area as a result. In this regard, not only is there occupational displacement, but also housing displacement.

Several academics have observed that there are two distinct types of displacement: direct and indirect (Davidson and Lees, 2005, 2010; Newman and Wyly, 2006; Slater, 2009; Davidson, 2010; DeVerteuil, 2011). While direct displacement refers to the emigration from the neighbourhood or the moment of eviction, indirect displacement is the most likely event in TEFL tourism, being a long-term process that results in a set of pressures that make it progressively difficult for low-income residents to remain over time. According to Davidson and Lees (2010), this aspect is extremely important in the understanding of displacement in contemporary gentrification. Indirect displacement has been found to have several long-term implications which affect host residents' quality of life on a day-to-day basis. Gentrification tends to be viewed as a long-term process in which direct displacement would be the last outcome. From this point of view, gentrification is not the moment when a community member has to leave his or her residence. Rather, a person feels gentrification from the very moment that different forces make it difficult or uneasy to continue living in the area. As a result, Davidson (2008, 2009) and Davidson and Lees (2010) suggest that the pressure of indirect displacement results in residents experiencing loss of place, where they are often forced into a new colonised social context.

Gentrification brings with it a number of further negative impacts on the community. There may be a loss of services and shops generally used by members of the host community which are substituted by amenities and consumption facilities designed to meet the needs of tourists. Areas that once offered affordable services are transformed by the presence of tourism and its associated activities, resulting in increasing affordability pressures placed upon host residents. Tourism expansion thereby influences new patterns of consumption (Davidson, 2010), which denotes a culture of displacement. This is particularly common in developing tourism regions where public space is privatised for tourism purposes. This exemplifies private ownership in the area and public areas that once were used as free communal facilities will often be transformed into establishments such as cafés, restaurants or festival marketplaces. These sets of indirect displacement pressures cannot be seen as independent forces, but as reciprocal elements that constrain the ability of low-income hosts and indigenous populations to remain in gentrifying neighbourhoods.

On another note, the fight against corruption needs to play a central role in the efforts to bring about sustainability in TEFL tourism, through the monitoring of individuals and organisations. Corruption is often an obstacle to development and comes in many different forms. Many big tour operators use tax havens and are associated with tax avoidance or even tax evasion. This is a problem for tourism which supposedly takes a 'pro-poor' stance, where money is retained by the TEFL organisation as opposed to being collected through the host country's taxation system. Corruption is often demonstrated in many countries through different forms of bribery, which causes major damage to societies by undermining justice and social participation.

While there is no 'quick fix' for economic issues such as those discussed above, there is certainly a place for corporate social responsibility (CSR). CSR is a method used by organisations to assess the impact that the company has on society, provoking them to consider putting policies in place to enable and support sustainable practices. Also referred to as corporate responsibility, corporate citizenship or responsible or

**Box 6.10.** Cheating in TEFL

Some might argue that the education system in Thailand has an element of corruption.

Cheating is rife.

A study by Stainton (2017b) found that many TEFL tourists were frustrated because of the levels of cheating in schools. This was largely attributed to the fact that in Thai culture they do not like to 'lose face' and therefore they do not like to fail tests or to get poor test results.

Cheating was not only undertaken by the children, but also by the teachers who were studying for higher qualifications alongside their teaching role.

There was evidence in this research that TEFL tourists lost motivation as a result, stating that the qualifications lost their value and worth.

sustainable business, CSR enables a business to assess its stakeholders, including employees, customers, suppliers, communities and the environment and to evaluate its responsibilities to each of them. Good CSR strategies will focus on creating policies which benefit society in a variety of different ways while taking the organisation's individual circumstances into consideration. Effective policies help to improve relationships with host communities, employees and customers. The overarching goal is sustainability, which complements business activities with socially responsible actions that support communities, individuals and the environment. While these aspects might not be at the forefront of the organisation's motivations, they can help to achieve its goals through enhancing its corporate image.

## Pedagogic impacts

The literature on TEFL tourism is distinctly polarised between tourism-based issues and aspects, and pedagogical practices. While the aim of this book is to examine TEFL tourism as a social concept, there is a place for pedagogy and this is an important part of the practice of TEFL. The difficulty lies in the complexity of the education industry, with cultural and geographical boundaries being important components. A Google search on the 'impacts of TEFL' shows results focused on one of two areas: benefits to the teacher (i.e. personal fulfilment/career development); or benefits to the host community (i.e. improved language). A Google search for 'negative impacts of TEFL' has results which are even more one-sided, with almost all hits focusing on the downsides of TEFL from the teacher perspective. Given the abundant discussions of good and bad teaching practices in a pedagogic sense, along with the extensive academic discussions of the negative impacts associated with tourism development, it is worrying that there is such little attention paid to this domain. While these areas were addressed in Chapter Five, it is important to re-emphasise said issues in the context of the impact of TEFL tourism.

There is an urgent need for the evaluation of accredited TEFL courses and qualifications. Does a 20-hour course qualify somebody to be a 'teacher'? What if their own levels of English are poor, but they completed a TEFL course? Is there any prerequisite for tourists to be familiar with aspects typically not covered in TEFL short courses such as product differentiation, administration processes, working with exam boards, inspectorate processes, etc.? These are questions for which there are currently no answers. Each destination, agency and educational institution seems to be interpreting its own rules which has contributed to an ambiguous TEFL tourism industry.

With so much global demand for English teachers, and with such a broad range of different TEFL, TESOL and TESL certificates, it can be difficult for organisations to decide what criteria they will put in place for the recruitment of their teachers. There are hundreds of different English teaching courses available, but without external validation there is no guarantee of the quality of service that the student receives or the calibre of their work. There is no way of knowing if the training provided will enable the TEFL tourist to do a good job, which can create difficulties for the TEFL agency or educational organisation in the long run. Without robust quality-assured training, teachers run the risk of working ignorant of best practices – put simply, unqualified teachers do not know what they do not know.

Unlike most educational subjects which are guided by content, English language education

**Box 6.11.** Paperwork, qualifications and legal requirements

A cause for concern identified in Stainton's (2017a) research was the legal working status among TEFL teachers. She found that 45.1% of TEFL tourists did not hold the correct legal documentation at some point during their TEFL placement in Thailand; 31.3% of these respondents blamed this on the agency or school which recruited them. Teachers without the legal paperwork were commonly on a gap year or backpacking trip (62.5%), were teaching in rural areas (48.8%), were unqualified (63.2%) and were under the age of 40 (78.2%).

This represents a proportion of the teaching community which, as a result of their lack of official paperwork and documentation, may not have been paying the correct taxes, be inappropriately qualified and may be unaccounted for by the state, thus undermining some of the economic and pedagogic benefits of TEFL tourism.

presents information about the language in English which requires students to apply skills, strategies and principles in order to change their own use of language. In effect, this means that the focus of lessons should not be on the teacher, but should instead be focused on the student and their outcomes. Defining clear lesson outcomes, developing staged instruction to work towards main and subsidiary lesson aims, planning activities to address specific aspects of language, with a view to student production of the language being taught, are sophisticated skills to achieve effectively, and it is unlikely that such skills will be developed while undertaking a short TEFL course over a period of hours, days or even weeks.

For most TEFL tourists it is actually necessary for them to 'unlearn' things before entering the classroom for the first time. Putting concepts into simple, layman's terms is not something that comes easy to many people, particularly if they have been encouraged to do exactly the opposite through academic writing, for instance. TEFL agencies have a responsibility to ensure that teachers are, at the very least, informed of classroom expectations and experiences. However, the agencies themselves are frequently unfamiliar with the practice of TEFL and are, therefore, not in a strong position to provide the necessary guidance and to implement the frameworks to ensure sustainable management of teaching practice.

Another serious concern within the TEFL industry, and more broadly within similar industries such as volunteer tourism or orphanage tourism, is the lack of background checks. Most western nations have their own systems for undertaking background searches when recruiting staff who will work with young or vulnerable people. In the UK, for instance, educational institutions require applicants to have a DBS check. Through working with the police, the Home Office, the Department for Education, the Department for Health and the Criminal Records Office (in the UK) are responsible for managing requests for criminal records checks (DBS checks) and deciding whether it is appropriate for a person to be placed on or removed from a barred list.

While this system may operate efficiently in the UK, along with similar systems in other western countries, there is little integration between them, meaning that somebody who is not deemed fit to work with children in the UK may be able to do so in other parts of the world because the intended destination does not hold rights to access the data held in the UK. This is especially the case in developing countries where the infrastructure for monitoring such issues is generally far less developed than it is in western societies. Indeed, it would be a difficult challenge to introduce an international database, whereby such relevant information can be stored and accessed across the world. Yet, it seems essential to share such protective measures since we are living in an increasingly interconnected world.

**Box 6.12.** Sexual exploitation in volunteer tourism

In her study of volunteer tourism perceptions, Wright (2014) found that there was evidence of sexual abuse in orphanages operating in Nepal. On undertaking interviews with members of the host community it was found that western men were volunteering to work in orphanages in the country, but were then exploiting their position of power by sexually abusing the children they were working with.

This is a significant concern, yet there is little research outlining the extent to which this may be taking place in Nepal and other countries around the world.

While a consistent monitoring approach would be an obvious advantage, consistency should not always be the aim in all areas of TEFL tourism. While globalisation is helping us move toward a world of 'sameness', which in many ways is viewed as a positive, it may not be sufficient to meet the needs of all people in all destinations. This is evidenced in the education system, where a child in rural Uganda may require a different approach to schooling than a Japanese student living in central Toyko. In fact, it can be argued that the increasingly standardised educational system is not suitable for all learners.

George Ritzer introduced the concept of 'McDonaldization' ' through his 1996 writings in the book entitled *The McDonaldization of Society*. He explains that the McDonaldization of society is a phenomenon which occurs when society, its institutions and its organisations have adapted to possess the same characteristics that are found in fast food chains, like McDonald's. This theoretical perspective builds upon sociologist Max Weber's theory of how scientific rationality produced bureaucracy (Weber, 1978). This became the central organising force of modern societies through much of the 20th century.

Weber posits that modern bureaucracy is defined by hierarchical roles, which are compartmentalised by knowledge and roles. A perceived merit-based system of employment and advancement and the legal-rationality authority of the rule of law are observed throughout many aspects of societies around the world. Ritzer (1996) believes that changes within science, economy and culture have shifted societies away from Weber's bureaucracy to a new social structure and order that he called 'McDonaldization'.

This new economic and social order is defined by four key aspects. The first focus is on efficiency, which consists of a managerial focus on minimising the time required to complete individual tasks as well as that required to complete the whole operation or process of production and distribution. The second aspect is that of calculability, where the focus is on quantifiable objectives (counting things) rather than subjective ones (evaluation of quality). The third is predictability and standardisation which encourages repetition and routine to facilitate consistent output of products or experiences. Lastly, control within McDonaldization is wielded by management to ensure that workers appear and act the same on a moment-to-moment and daily basis.

Ritzer ascertains that these four characteristics are not only observable in production, work and within the consumer experience, but that their presence ripples through all aspects of social life. McDonaldization in turn affects the values, preferences, goals and worldviews of a society along with unique identities and social relationships. McDonaldization is in fact a global phenomenon, which has been driven by corporations who hold the economic power and cultural dominance of the West, which leads to a global homogenisation of economic and social life.

While there is little evidence to suggest that the TEFL industry has become 'McDonaldized' in this way, this theoretical standpoint does provide an indication of the possible future direction of the industry. The fourth aspect of control may be difficult to achieve given the inherent staff changes and different backgrounds among TEFL tourists, but there is evidence that TEFL tourism is likely to create a dilution in individualism, paying less attention to the unique and specific needs of the community it serves and more to globalised views of what education has become in the eyes of consumerism.

Education has, in many societies, become a commodity. In the UK for instance, the marketisation of higher education, with profit-making colleges and debt-laden customers has resulted in a profit-driven marketplace that does not necessarily focus on the needs of the student, but instead is based largely on market influences and profit incentives. This can be seen in many ways as being parallel to the TEFL tourism industry, where the focus has again moved away from the ideals of community development and philanthropy towards consumerism and meeting the demands of the capitalist marketplace.

Throughout the past decade, a succession of steps have been taken to consolidate the marketisation of higher education, particularly in the UK. As in TEFL tourism, it is relatively easy for private providers, both for-profit and not-for-profit, to obtain qualification-awarding powers. Also like TEFL, there is no longer a cap on student numbers, which incentivises institutions to maximise their income by admitting more students. The result has been a change in the character and, above all, the ethos of educational institutions. In the UK, universities are now forced to regard each other as competitors in the same market, and their success is dependent on the accuracy with which they pitch their products to appeal to their particular niche of consumers. 'Value for money' has become the new 'buzz word' in academia and there is a growing expectation by students that because they have paid money for their courses they should be awarded the desired outcomes (i.e. the qualification with a good mark). Universities are choosing not to spend their income on much-needed educational resources, but instead on tangible aspects of the student experience that appeal to the consumer such as lavish halls of residence, in much the same way that a TEFL package may emphasise accommodation near the beach and tourist excursions.

While there appears to be no research into the effects of marketisation in the TEFL sector, there is ample evidence of the effects of marketisation within higher education. There is now a completely different attitude to admission. In the UK, universities had always been regarded as selective: from among those who applied for admission, universities chose those judged best able to cope with the level of education being offered. Those deemed worthy of the university place were offered one and others, even those with good grades, were turned away. Nowadays, admission is not a matter of selection, but rather of recruiting. Institutions are hungry for the fees that students bring and are anxious to take as many as possible. This has resulted in entry requirements being lowered and the standards of students dropping, particularly among lower-ranked universities. Just as there is a concern regarding the calibre of TEFL tourist graduates and the quality of their output, there is a growing concern that the standard of university graduates is also in decline as a direct result of marketisation.

Although it does not solve many of the problems associated with marketisation, the UK is trying to manage this newly developed capitalist marketplace through the introduction of the Teaching and Excellence Framework (TEF). The intention is that this method of 'grading' universities will in time force a differentiation of fees between institutions, resulting in the differential pricing of courses. The focus is on the concept of 'value for money', where students will pay more or less depending on the average outcomes deriving from previous students of that particular institution or course. This may result in reforms such as the removal of loans for those courses with an unsatisfactory record of post-graduation employment; a greater say for employers in what is taught; and two-year degrees and more private providers, with their students being eligible for loans on exactly the same basis as those at 'public' universities. Under the forces of such a competitive marketplace it is likely that at some point an institution will be declared on the edge of bankruptcy and forced either to close or to merge with another institution. There will definitely be a sharp rise in student complaints and very probably a big increase in litigation against universities.

While such a system may have its pitfalls, it is the consequence of commercialising and commodifying education. If the student is viewed as a consumer, there is a strong argument that they should have the rights of a consumer. Yet, while this is being recognised within the UK higher education system, which shares many parallels with the commercialisation of TEFL within tourism, there is little recognition of this within the TEFL sector. The 'grading' of TEFL qualification providers may not necessarily be the answer, but there is most certainly a need for some kind of procedure or practice allowing stakeholders to identify the worth and value of the qualifications obtained.

**Box 6.13.** The Teaching and Excellence Framework (TEF)

The TEF is a scheme recently introduced by the government to measure the quality of teaching at Higher Education Institutions (HEIs) in England. Universities are given a rating to indicate the level of teaching quality that they provide. In future years it is intended that this rating will be awarded not just at an institutional level, but also at a discipline level.

The scheme is administered by the Higher Education Funding Council for England (HEFCE). They support in delivering TEF by staff at the Quality Assurance Agency for Higher Education (QAA). Once established the Office for Students (OfS) will take over responsibility for the implementation of TEF from HEFCE.

The government's stated intention is that TEF will be used

1) to 'provide clear information to students about where the best provision can be found'
2) to 'encourage providers to improve teaching quality to reduce variability' and
3) to 'help drive UK productivity by ensuring a better match of graduate skills with the needs of employers and the economy'

TEF will also be linked to the ability of a university to raise the level of tuition fees charged to a student.

(Political Studies Association, 2018)

## Ethics within TEFL Tourism

This chapter has demonstrated the possible positive and negative impacts of tourism. Much of this can be summed up as ethical practice, an imperative notion underlying the concept of sustainable tourism. The concept of ethics is based on perceptions of what is right and what is wrong. Ethical standards prescribe what humans should do, usually in regard to rights, obligations, benefits to society, fairness or specific virtues. Ethical beliefs do in fact shape the way in which society lives. Ethical questions explore what Aristotle called 'a life well-lived'.

Complex ethical problems, both within TEFL tourism and more broadly, can be individual and private or widespread and systemic, involving groups, organisations or whole communities. Ethics provides a framework from which stakeholders can operate effectively, allowing agents of TEFL tourism to be consistent in their judgements, provide reasons for beliefs and to critically examine practices and operations. Most importantly, ethics allows a set of core values and principles to be developed and worked within. Ethical people have what philosopher Thomas Aquinas called a 'well-informed conscience'. They live what Socrates called 'an examined life', which is defined as a life particularly associated with being human. Ethical people try to answer the question of how to live by reflecting on difficult situations. They then act in a way that is true to who they are and what they believe.

From a business sense, ethical practice can attract customers to the organisation's products, helping to increase sales and profits. It can reduce staff turnover if both TEFL agency staff and teacher tourists as employees want to stay with the business, which therefore increases productivity. It can also attract more employees wanting to work for the business, reduce recruitment costs and enable the company to get the most talented TEFL tourists. Lastly, ethical practice helps to attract investors and keep the company's share price high, thereby protecting the business.

But ethics within TEFL tourism permeates beyond business priorities. It is about suiting the needs of the community, the TEFL tourist and the TEFL agencies along with any wider-arching incentives intended by governments or other stakeholders. There is, of course, an inherent paradox between capitalism and philanthropy, and that has been discussed within this chapter. The important thing is to manage this effectively to capitalise on the benefits of TEFL tourism for all stakeholders and to reduce the negative impacts noted in this chapter among other possibilities.

# 7

# The TEFL Tourist – Who, What and Why?

Given that TEFL tourism is a new industry, inevitably little is known about it, including those without whom the industry would cease to exist: the tourists. It would be inappropriate to generalise all TEFL tourists across all localities – there are so many variables that contribute to the general make-up of the industry, ranging from motivational factors to characteristics. This chapter therefore discusses the premise of one case study which was based on TEFL tourists in Thailand (Stainton, 2018b). While the proceeding conclusions have not been tested across multiple destinations to date, there is considerable potential for generalisation in some areas, particularly in destinations which experience similar numbers of TEFL tourists in similar geographical areas and have similar cultures or complementary education systems.

## TEFL Tourist Characteristics

There are a number of common characteristics that can be identified between TEFL tourists. In general, TEFL tourists tend to be relatively young, with the majority aged between 22 and 30 (Stainton, 2018b). This is likely to be the result of motivations, where many may choose to undertake a TEFL placement in order to gain work experience and to enhance their CV. It may also be due to the popularity of gap years among people of this age. This average age is notably higher than the average age to take part in volunteer tourism, where participants tend to be between the ages of 18 and 22 (Wearing, 2001; TRAM, 2008; Tomazos and Butler, 2009). This could be because many TEFL placements are paid, enabling tourists to remain in employment for a longer period of time and possibly eliminating the need for additional funds, where the TEFL employment can satisfy the tourist's financial requirements. It may also be the result of the case study from which this data was collected: in Thailand there is a large population of retirement-aged men who operate as TEFL teachers and therefore this may skew the data in this direction.

As with volunteer tourists (Gecko *et al.*, 2009; Tomazos and Butler, 2009), the majority of TEFL tourists are single. Relationship status tends to differ with age and it is logical to conclude that the majority of young TEFL tourists are unlikely to be in a serious relationship.

TEFL tourists mainly come from five dominant nationalities: British, American, South African, Irish and Australian (Stainton, 2017a, b, c, d). This correlates with the accepted nationalities indicated on many of the TEFL recruitment agency websites. There may be some differences globally between nationalities in particular locations as a result of proximity. For example, there may be more Americans in Mexico and Costa Rica because it is close to their home country, whereas Fiji and Indonesia may attract more

**Box 7.1.** Ancestral tourism

Recent years have seen a significant rise in the number of people who wish to find out more about the history of their family, their home and their community.

Ancestral connections are a significant motivator for tourism activity and it is suggested that people making ancestral journeys tend to stay longer, spend more and prefer to consume local products. Ancestral tourism is an important, but often under-represented, part of the tourism economy in many destinations.

There are particular destinations that are popular for ancestral tourism, such as parts of Africa which were previous slave destinations or India where there has been extensive economic migration over the years. These are destinations that are likely to have greater appeal than others to TEFL tourists who have a background or history in these regions.

**Box 7.2.** Illegal TEFL tourists in Thailand

As noted in Box 2.5 (p. 15) all foreign teachers need a non-immigrant B visa and must present their TEFL certificates, undergo a criminal record check and present an official letter of recommendation from the Thai education institution where they will be based. Information about their contract or term of employment is also required. Similarly, volunteers must obtain a non-immigrant O visa by providing official recommendation from the organisation in Thailand at which they are based and information on the term of voluntary work (Royal Thai Embassy, 2016).

Numerous respondents, however, do not meet the legal requirements (almost half). In support of Methanonpphakhun and Deocampo's (2016) findings, many respondents in Stainton's (2018b) research indicated that the process was complicated and the most common reasons for illegal employment were paperwork delays and lack of organisation by the school or agency responsible.

Australians. There may also be links to former colonialism, where for instance India may have more people from the UK, whereas Ireland has a diaspora throughout the western world.

The majority of TEFL tourists are qualified to degree level or above. Again, this is in line with volunteer tourism (Tomazos and Butler, 2009; Birdwell, 2011). Stainton (2018b) found that a quarter of TEFL tourists were continuing students, almost half of whom were studying teaching-focused qualifications. This makes sense given that many TEFL tourists are young (the most popular time to study) and that many of them are undertaking their TEFL placement in order to gain experience and to enhance their CVs. It is also logical given that a TEFL qualification and programme of study is a core element of the TEFL package, as discussed in Chapter Four.

As touched upon in the previous chapter, a significant number of TEFL tourists do not hold the correct legal documentation to be living and working in the location in which they are based. This is a serious concern for the industry. It makes it impossible to be able to examine the size, scope and impacts of the TEFL industry when a proportion of the teaching community are 'hidden' through illegal employment (Punthumasen, 2007; Kirkpatrick, 2012). It also makes it extremely difficult to collect the relevant employment taxes and monies that should be raised through visa fees, etc. This reduces the benefits that could be brought to the destination from TEFL tourism and undermines some of the positive economic impacts from employing tourists in this way.

There is an argument to say that many TEFL tourists do not have the required skillset. It is common for TEFL tourists to be 'out of field', meaning that they do not hold qualifications relevant to teaching or the subject being taught, which has been known to lead to poor quality of work undertaken (Guttentag, 2009; Benson and Wearing, 2012). This is an issue that has been discussed in this book and raised in the study of TEFL tourists in Thailand. The limited skill set possessed by TEFL tourists was further emphasised by a lack of previous experience – under half of TEFL tourists had taught before. Unlike volunteer tourism, however, where not all roles require specialisms, teaching skills and subject knowledge are integral to the quality of the education provided by the TEFL tourist.

As with research focusing on the lengths of volunteer tourism placements (Callanan and Thomas, 2005; Gecko *et al.*, 2009; Keese, 2011), there was a strong variance between the duration of TEFL tourism. In support of the figures presented by the TESOL International Association (2014), which found that 50% of TEFL teachers teach for up to one year, this research found that almost half of TEFL tourists teach for a duration between one week and one year and the most common length of a TEFL placement is one to two years. Most TEFL tourists tend to be located in major cities or rural areas, although there are opportunities available in desirable tourist areas such as beach resorts, but these are not as easy to come by.

As previously noted, the main difference between TEFL tourism and volunteer tourism is that the majority of TEFL tourists are paid for their services. Official salaries in Thailand ranged from £70–£2,775 per month, with an average salary of approximately £700 (Stainton, 2016). Many TEFL tourists are also entitled to expenses which include free accommodation or a housing allowance, food, travel reimbursement, visas and work permits and insurance. Salaries differ between TEFL destinations depending on the cost of living in the area, although they do generally appear to be enough to cover living costs with a small surplus.

On average, TEFL tourists taught for 16–20 hours each week in a range of educational institution types categorised as government-funded schools (64.2%), privately-funded schools (45%) and language institutions (22.3%). Secondary level teaching was the most common among TEFL teachers in Thailand, closely followed by junior level (30.5%), infants (14.1%), preschool (11.8%), university (9.9%) and adult classes (6.3%). Over half of the respondents (54.9%) indicated that they taught across more than one student level.

In Thailand, almost half of TEFL tourists organise their TEFL placement via a TEFL agency, outlined in Chapter Four. For those who paid an initial fee to an agency, the average amount was approximately £550. This is significantly lower than the average cost of £2,000 to undertake a volunteer tourism trip (TRAM, 2008). In contrast to the volunteer tourism sector, however, a quarter of TEFL tourists do not pay an official initial fee; instead the agency takes a percentage of their salary each month, although the exact amounts are not known. Some TEFL tourists are not even aware that they pay an agency fee as it is taken directly from their salary.

## TEFL Tourism Motivations

Tourist motivation is an important area of tourism management that continues to be addressed by researchers. At a simple level, it seems obvious that people do things because they either need or want to. People go to work, for example,

**Box 7.3.** Why be a TEFL Teacher in Hong Kong with i-to-i?

From the stark beauty of the Chinese coastline to the neon cityscapes; there's nothing quite like teaching English in Hong Kong.

Plus, with English now one of the official languages of Hong Kong and near essential to finding a well-paid job for Hong Kong citizens, you can see why there is such a high demand for English teachers from overseas.

Since its colonization by the British back in 1843, Hong Kong has been a major player on the global scene and is home to over half a million expats. This internationalisation of Hong Kong has meant that the cost of living here is quite a lot higher than in Mainland China. Nevertheless, it is still possible to live comfortably on a teacher's wage, which is on average £1500 a month.

As land in Hong Kong is scarce, accommodation is going to be your biggest outlay. In the trendy Hong Kong Island, you will pay around £700 for a studio apartment, whereas in Kowloon or the New Territories you will pay around 20% less.

Not everything in Hong Kong is expensive though and eating out is surprisingly cheap. A typical Cantonese meal of dim sum or egg tart will cost you only £3/$5.

Another thing that you'll be able to do cheaply in Hong Kong is to go out for a few drinks (particularly if you're female). It's common for Chinese bars to host ladies' nights where women drink for free all evening in an attempt to entice the male locals!

(i-to-i, 2017)

in order to be able to afford to feed their families whereas they go on holiday in search of enjoyment or satisfaction. Motivation is, however, far more complex than the simple concepts of need or want. There are many other variables and factors which play key roles in motivation. Many theorists have tried to investigate these, although it remains a contested subject.

Overall, the basic perspective on motivation tends to involve the three concepts of needs, behaviour and satisfaction. To put it simply, people have certain needs or wants, and this causes them to do certain things (behaviour), which satisfy those needs (satisfaction). Motivation for a need, such as earning a living, are typically seen as very different from motivation for pleasure, such as taking a holiday. TEFL tourism is unique in that it involves aspects of both work and leisure, which inevitably makes for a complex analysis of motivation.

Many motivation theories posit a hierarchy of needs, suggesting that the needs at the bottom are the most urgent and must be met before attention can be paid to the others. Maslow's (1943) hierarchy of needs is the most famous example (Fig. 7.1).

According to Maslow's original theory of human motivation, one progresses through the hierarchy from 1 to 5. This is, however, widely contested and Maslow himself later proposed that the order in the hierarchy 'is not nearly as rigid' (p. 68) as he may have implied in his earlier description and he extended the idea to include his observations of innate curiosity. A TEFL tourist, for instance, is unlikely to say 'I can't socialise with my friends this evening because I haven't had sex in three days!' This demonstrates that not all physiological needs have to be satisfied before higher needs. Maslow's model can also be critiqued with regard to culture, a particularly relevant issue when studying people deriving from and working within a range of differing cultural contexts as found in TEFL tourism. It can be argued that 'lofty' goals like personal growth and creativity at the top are not desired or seen as important in some cultures, particularly in developing countries. Whereas for some, 'base' instincts like sex are not fundamental to their motivation, particularly in areas where people are getting married and having children later in life and where sexual relationships are not seen as being as important as they may be in other cultures. There is also the question of whether self-actualisation is actually a fundamental need or whether it is better described as just something that can be done in one's leisure time.

Alderfer (1969) classifies needs into hierarchical categories in his ERG theory (Fig. 7.2). Consisting of growth needs, relatedness needs and existence needs, his theory is very similar to Maslow, although perhaps a bit more rational. For example, in Alderfer's model, sex does not need to be in the bottom category as it is in Maslow's model, since it is not crucial to (the individual's) existence. So by moving sex, this theory does not predict that TEFL tourists have to have sex before they can think about going to class or socialising, as Maslow's theory does.

Cognitive evaluation theory (CET) is another approach that some scholars and industry

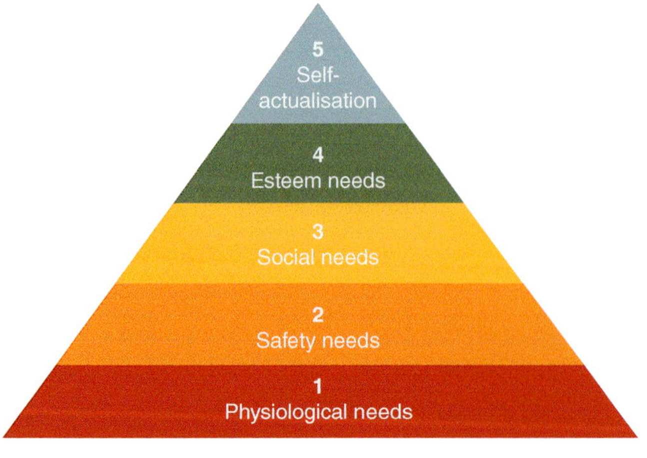


**Fig. 7.1.** Maslow's hierarchy of needs

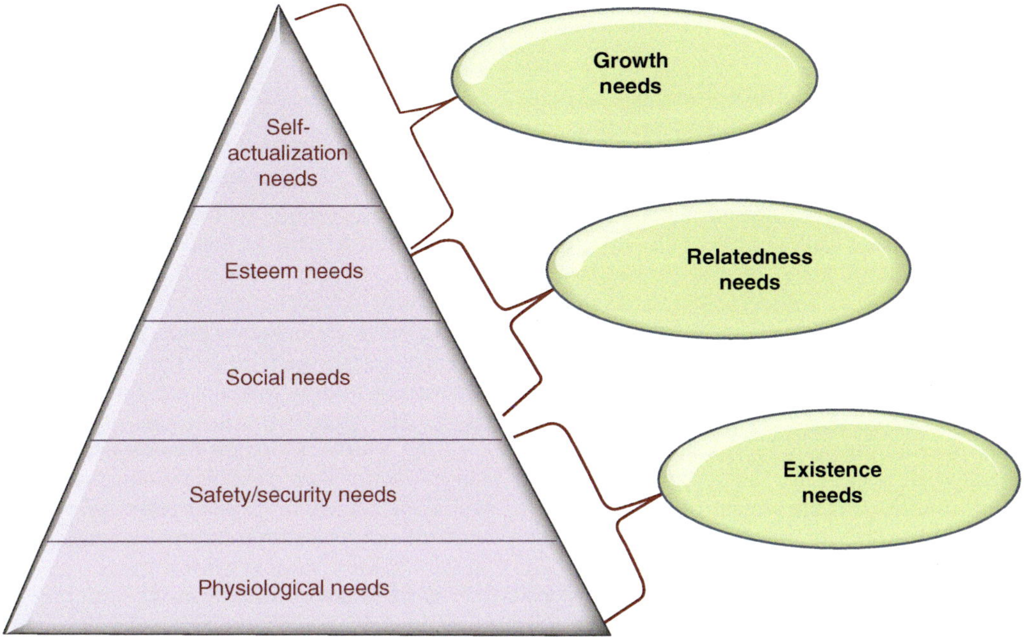


**Fig. 7.2.** Alderfer's ERG theory

practitioners take to assess motivation. This theory suggests that there are actually two motivation systems: intrinsic and extrinsic. Intrinsic motivators include areas such as achievement, responsibility and competence. Motivators come from the actual performance of the task or job and from the intrinsic interest of the work. Extrinsic motivators, on the other hand, include areas such as pay, promotion, feedback and working conditions. In other words, they come from a person's environment that is controlled by others.

Intrinsically motivated people typically perform for their own achievement and satisfaction. If they come to believe that they are doing something because of the pay or the working conditions or some other extrinsic reason, they begin to lose motivation, which is likely to affect the quality of work undertaken. The belief is that the presence of powerful extrinsic motivators can actually reduce a person's intrinsic motivation, particularly if the extrinsic motivators are perceived by the person to be controlled by other people. In the context of TEFL tourism, this could mean that those motivated by salary or by enhancing their CV may perform less well than those who are philanthropically motivated or who do it for enjoyment. Specific motivations in this area are discussed shortly.

According to Herzberg *et al.* (1959), two kinds of factor affect motivation. Hygiene factors are factors whose absence motivates, but whose presence has no perceived effect. They are things that when taken away, people become dissatisfied and act to have them back. Take working conditions, for example. Western TEFL tourists are likely to make an assumption, or perhaps not even give much thought to the fact that working conditions will be the same in their TEFL destination. But the reality is that they may not. They may have to work in high temperatures with no air conditioning; they may not be paid sick leave or have an adequate training programme. These are aspects that the TEFL tourist may not actively seek, but that they desire once they no longer have them. In general, hygiene factors are extrinsic items low in the Maslow/Alderfer hierarchy.

Motivators, on the other hand, are factors whose presence motivates. Their absence does not cause any particular dissatisfaction, it just fails to motivate. In terms of TEFL tourism, this could include working in a sunny destination, having access to many touristic activities, developing new social bonds, having a feeling of 'helping others', etc.

Equity theory is another approach to the study of motivation and is one which is particularly relevant to TEFL tourism given that frequently there are disparities between local and foreign salaries, as noted in Chapter Six. Adams' (1963) equity theory calls for a fair balance to be struck between an employee's inputs (hard work, skill level, acceptance, enthusiasm, etc.) and an employee's outputs (salary, benefits, leisure time, intangibles such as recognition, etc.).

According to the theory, finding this fair balance serves to ensure that a strong and productive relationship is achieved with the TEFL tourist, with the overall result being contented, motivated employees, TEFL agencies, school staff and those being educated. Higher motivation is also likely to yield better results in terms of quality of work and associated outputs such as exam grades. There is an important discussion to be had about fairness, and whether it is morally correct for TEFL tourists to be paid different salaries or for their salaries to differ from those of qualified teachers in the host destination. In fact, it is possible that TEFL tourists, while being well paid, may actually feel demotivated as a result of the hostility caused by the differences in salaries between them and the locals. This is an area warranting further research.

Reinforcement theory of motivation was proposed by Skinner in 1948 who claimed that a person's behaviour is a function of their consequences. The theory is based around the concept of 'law of effect' – i.e a person's behaviour with positive consequences is likely to be repeated, whereas a person's behaviour which has negative consequences is unlikely to be repeated. This theory overlooks the internal state of a person – i.e. their inner feelings, which are ignored by Skinner. Instead the theory focuses explicitly on what happens to a person when they take some action. In essence, the external environment of the organisation must be designed effectively and positively so as to motivate the employee.

There are four prominent methods for controlling the behaviour of employees (or TEFL tourists). Positive reinforcement implies giving a positive response when a person demonstrates positive behaviour. For example, a manager may immediately praise a TEFL tourist for producing good exam results. This increases the likelihood of outstanding behaviour occurring again. If the TEFL tourist's behaviour improves, rewards can be a positive reinforcer. In contrast, negative reinforcement tends to involve rewarding an employee through the removal of negative or undesirable consequences. Both positive and negative reinforcement can be used for increasing desirable or required behaviour. Punishment typically involves the removal of positive consequences in order to lower the probability of undesirable behaviour being repeated in the future. Lastly, extinction implies the absence of reinforcements, which lowers the probability of undesired behaviour through the removal of rewards. This could be demonstrated, for instance, if a TEFL tourist no longer receives praise and admiration for their good work. They may feel that their behaviour is generating no fruitful consequence as a result. Extinction may unintentionally lower desirable behaviour.

The last theory worthy of mention here is expectancy theory. This is designed to bring together many of the elements of the previous theories, combining the perceptual aspects of equity theory with the behavioural aspects of others. Basically, it proposes that motivation is the result of the concepts of expectancy, instrumentality and valence. According to this theory, motivation is a function of the following.

Expectancy is the TEFL tourist's perception that effort will result in performance – i.e. the tourist's assessment of the degree to which hard work correlates with performance. Essentially it is the tourist's perception that their performance will be rewarded/punished – i.e. the tourist's assessment of how well the amount of reward correlates with the quality of performance. Valence is the perceived strength of the reward or punishment that will result from the performance. If the reward is small, the motivation will be small, even if expectancy and instrumentality are both high.

TEFL tourism motivation is a large topic by itself. However, by outlining the commonly noted theories of motivation, there is scope for these to be applied and assessed within the context of TEFL tourism in future research. The proceeding paragraphs outline the TEFL tourist motivations identified within Stainton's (2018b) research on Thailand.

Key motivational themes for becoming a TEFL tourist can be broadly grouped into aspects relating to travel and tourism, life experiences such as skill development or new experiences, and destination-specific motivations such as Thai food or culture.

Over half of TEFL tourists surveyed strongly agreed that travel and tourism was a motivating factor, thus further emphasising the links between TEFL and tourism. For some, TEFL teaching was secondary to the travel experience, where it is used as a mechanism to fund their travels. This notion is supported by Methanonpphakhun and Deocampo (2016), who found in their research that most foreign teachers were not motivated as a result of their love of teaching, but rather to travel, for education or to start a new life. Furthermore, Punthumasen (2007) described foreign TEFL teachers in Thailand not as teachers, but as tourists, who bring along the negative connotations associated with lack of relevant experience, skills and qualifications.

**Table 7.1.** TEFL tourism motivations in Thailand

| Motivation | % of TEFL tourists |
|---|---|
| Wanted to live/work abroad | 87.8 |
| Wanted to experience the Thai culture | 85.4 |
| Wanted to do something different | 81.1 |
| Craved adventure | 81.0 |
| Wanted to travel | 74.1 |
| Develop teaching skills | 64.9 |
| Learn more about themselves | 69.7 |
| To do something good | 64.7 |
| See if teaching is right for them | 63.3 |
| The weather was appealing | 61.9 |
| The cost of living was cheaper | 61.7 |
| Enhance CV | 57.7 |
| For a slower pace of life | 45.3 |
| Didn't know what else to do with life | 44.3 |
| The food was appealing | 42.5 |
| It was recommended | 36.7 |
| Was easier to find a job in Thailand than at home | 34.7 |
| Access to beaches | 22.6 |
| The nightlife | 20.1 |
| Was in a romantic relationship with a native | 15.9 |
| It was the path God intended | 13.4 |
| Part of an educational course | 9.7 |
| Sex tourism-nightlife | 5.6 |

(Stainton 2018b)

There are many motivations which can be situated beneath the travel and tourism umbrella. Motivations exhibited by respondents such as having the opportunity to visit the beach and to become involved with nightlife activities demonstrate clear links with the common sun, sea and sand motivations frequently associated with package holidays (Poon, 1993) and nightlife tourism (Calafat *et al.*, 2010; Tutenges, 2013). While these elements are not necessarily applicable to all TEFL tourism destinations it is important that when coupled with aspects associated with mass tourism, this phenomenon is recognised and managed in the appropriate way.

Some may question whether TEFL teacher motivations based on tourism elements are 'noble' or 'right'. Research in the volunteer tourism sector has categorised tourist-based or otherwise motivations through the development of various typologies. It has been suggested that volunteer tourists be categorised as vacation or volunteer-minded (Brown and Morrison, 2003) and a VOLUNtourist or volunTOURIST (Daldeniz and Hampton, 2011). While a similar approach, where TEFL teachers are categorsied as teacher-orientated or tourism-orientated, may provide an indication of the TEFL teacher's motives, it would fail to take into account the complexities and range of motivations identified (Fig. 7.1). Furthermore, it can be questioned whether such a categorisation is necessary, as selfish motives may not necessarily result in the positive outcomes of TEFL teaching. It is also important to note that TEFL teachers can possess any number of motivations, some of which may be more prominent than others.

Many TEFL tourists are also motivated by cultural integration. Those motivated to become a TEFL teacher in Thailand because they want to immerse themselves into the Thai culture demonstrate links with cultural tourism, whether this be through the consumption of products such as sites or monuments (Bonink, 1992; Munsters, 1994), or experiencing cultural aspects such as religion, ethnic traditions, linguistics and learning about Thai heritage through community engagement (MacIntosh and Goeldner, 1986).

Motivations thematically classified as life experiences, defined as any endeavour which facilitates experiences that the TEFL teacher may not have otherwise had, tend to be akin to the concept of lifelong learning, fostering the

**Box 7.4.** Top reasons to visit Thailand

**1.** Affordable food and lodging
Money goes a long way in Thailand. There are hundreds of bargain hotels and guesthouses. Thai food is likewise inexpensive. Thailand also has some outstanding hotels and resorts, all for significantly less than you would pay back home.

**2.** The weather
Whether you visit in December or May, expect warm weather and mostly clear days.

**3.** Beautiful beaches
Thailand has some of the best beaches in the world: warm, clear water; stunning vistas; and marine life to gawk at as you swim, snorkel or scuba. Plus there are so many different types of beaches in Thailand, from quiet and laid back to party all night, that everyone can find the perfect fit.

**4.** Terrific Thai food
You'll easily find favourites such as pad thai and curry, but you'll also find a wide selection of salads, noodles and snacks.

**5.** Shopping
If you like markets, bargains, or cool and quirky stuff, you'll find shopping in Thailand rewarding. Big city markets carry everything from shoes to art, and the malls in Bangkok are some of the fanciest in the world.

**6.** Easy travel
Once you're in the country, it's easy and cheap to get around, thanks mostly to local budget airlines like Nok Air, Air Asia and Bangkok Airways. The trains, boats and buses are also pretty good.

**7.** Interesting cities
The booming, bustling capital of Bangkok and Chiang Mai are very popular amongst tourists with lots of activities and attractions.

**8.** The culture
They call Thailand the 'Land of Smiles'. While there are occasional tourist scams and once in a while, you'll encounter someone who isn't smiling, by and large, the people of Thailand are friendly, open, and kind to foreign visitors.

**9.** Buddhist temples
Buddhism is Thailand's predominant religion, and it influences everything from the architecture to the way people interact with one another. All of the country's major cities have breathtaking Buddhist temples to visit.

**10.** Nightlife
There are go-go bars, discos, nightclubs and regular bars – something for every nightlife-lover.

continuous development and improvement of knowledge and skills needed for employment and personal fulfilment throughout life (Broomhall *et al.*, 2010; Falk *et al.*, 2012). This includes motivations to enhance a CV (58%), develop teaching skills (65%) and trial teaching as a career (63%), demonstrating connections with educational tourism (Ritchie *et al.*, 2003). Prominent motivations to aid skill development have also been demonstrated in a number of studies addressing the volunteer tourism industry (e.g. Soderman and Snead, 2008; Coghlan and Fennell, 2009; Daldeniz and Hampton, 2011), although these tend to be generic skills, as opposed to the specific teaching skills acquired through the practice of TEFL. A further 70% were motivated to learn more about themselves.

All of the motivations discussed so far in this chapter are largely hedonistic in essence, where the pursuit of pleasure and self-indulgence, whether through travel satisfaction, educational benefits or otherwise, permeate the stated motivations. While ethnocentric assumptions may lead to the belief that TEFL teachers are prominently driven by philanthropic motivations, there is no evidence of this to date. Although the majority of respondents in Stainton's (2018b) research identified that doing something good was a motive, as suggested by Lyons and Wearing (2012), this may not necessarily be indicative of their primary motivations. While it is important to remember that TEFL teachers can hold multiple motivations simultaneously, whether altruistic or selfish, (Tomazos and Butler, 2010; Hustinx, 2001), it has also been suggested that there is no such thing as an unselfish tourist. In fact, Wearing and McGehee (2013) claim that volunteer tourists are likely to disguise their selfish motivations behind their altruistic facade.

While there are some common elements between destinations to an extent, it is important to consider TEFL tourism with regard to the specific host destination. Motivations such as the reduced cost of living, a slower pace of life and the ability to easily secure employment can be attributed as destination-specific. Thailand is home to a sizeable sex tourism industry, encompassing both casual endeavours such as prostitution or sex shows (Kusy, 1999) and opportunities for long-term romantic engagements (Ruenkaew, 1999; Sims, 2012). For some, these lifestyle choices are motives for becoming a TEFL teacher in Thailand; sex tourism, nightlife opportunities and prospects of a romantic relationship with a native Thai are common occurrences.

## TEFL Typology

Within tourism literature there are a large number of tourist typologies, many of which have been proposed to allow for the subdivision of tourists into homogenous groups. The primary intention of these typologies is to effectively target and position organisational strategies, marketing operations, pricing, etc. Defining criteria typically include aspects such as demographics (age, income, education level, etc.), family characteristics, the tourist's predisposition towards benefits and their behaviour, characteristics and motivations. Geographic and economic criteria have also been suggested and used (Swarbrooke and Horner, 2007). Some of the most important criteria that have been proposed in typologies are the sociopsychological, as is the typology proposed in this book for a typical TEFL tourist (Fig 7.3). Such models attempt to provide a complete 'image' of the tourist. The criteria that are used in these cases tend to be focused around values and lifestyle, attitudes, interests and opinions, motives of travel and personality types. The most important typologies are those of Plog (1987) and Cohen (1972).

The TEFL tourist typology presented in this book is fundamentally based upon a variety that have been developed in tourism research, refocused to TEFL tourism. As noted above, Cohen was one of the first pioneers to attempt to typologise tourists and in his 1972 journal article 'Towards a Sociology of International Tourism', he first suggested that there are different types of tourists. Cohen's model is an early typological framework which is aimed at the sociological analysis of the tourism industry through the identification and analysis of four different categories of tourists. Cohen sought to define the continuum of possible combinations of familiarity and novelty sought by different tourist types in terms of their choice of tourist business establishment and geographic destination.

The two institutionalised forms of tourism identified by Cohen encompass the organised mass tourists and the individual mass tourists. The former are highly dependent on the travel industry's infrastructure during their holidays, with little requirement for organisation on their part. These tourists consume packaged trips at competitive prices, often for destinations made familiar through mass advertising and previous visits. They often travel in large groups with other tourists where the itineraries are inherently inflexible and predetermined. Cohen also identifies the individual mass tourists who are partly dependent on the tourism industry and its infrastructure, consuming most of the elements of the packaged tour (such as travel and accommodation), but also planning and structuring parts of their holiday themselves. This tourist type presents some synergy with the TEFL tourist who organises their experience via a TEFL operator, where essential elements such as accommodation and a TEFL course may be arranged for them, but they use their leisure time to do as they wish.

Cohen's (1972) non-institutionalised tourists are the opposite of the institutionalised tourists and are classified as either explorers or drifters. The explorers will arrange their travels consciously and independently, away from the tourist industry and its elements. They seek novelty beyond the mass generic tourist experience and often travel with specific goals in

**Box 7.5.** The TEFL 'traveller'

Jonny, owner of the blog 'One Step 4Ward' is famous for his travels around the world. He has now visited every country in the world and his journey started with his job as a TEFL teacher.

He states that he 'left Ireland in 2006 broke, taught English in Thailand and Korea, travelled A LOT, worked in Australia, started my blog, monetized, started more blogs, made $1m+ from my laptop, visiting every country in the world ... '

Jonny is an example of a leisure-minded TEFL tourist with elements of an expatriate-minded TEFL tourist. His TEFL experiences enabled him to fund his travels, at least in the beginning, and to live the lifestyle that he wanted to live in both Thailand and Korea.

This is typical of a long-term traveller and is a common motivation for entering into the TEFL marketplace.

One Step 4Ward (2018)

mind, such as relaxation, discovery, education and entertainment. They typically have more contact with the host community than the mass tourist, although there is no complete immersion. Taking the role of cultural immersion one step further, the drifters are more like wanderers, not considering themselves as tourists in the conventional sense. These tourists will arrange their trips alone with an aversion for conventional tourist places, infrastructure, arrangements and groups. They seek the excitement of complete strangeness and direct contact with new and different people. These individuals typically seek little comfort and are the ones who will hitchhike, ride a motorcycle, eat from street vendors and buy ingredients to cook their own meals. Drifters often travel inexpensively and might work to support themselves. According to Cohen, drifters tend to be children of affluence; often students or graduates who are not yet on the job market, seeking experiences completely different from those of their comfortable middle-class lives. TEFL tourists are also likely to fall into the category of non-institutionalised tourists, particularly if they organise the individual elements of their TEFL trip themselves. Aspects such as the inherent cultural immersion experiences when working in a foreign classroom sit well with this tourist type.

One of the best-known models in the travel and tourism field was developed by Plog in 1987, and it has been used as a reference for planning and designing tourism infrastructure and marketing projects for many years. A total of 16 domestic and foreign airlines, airframe manufacturers and various magazines sponsored Plog's research over a number of years in order to better comprehend the psychology of certain groups of tourists. The purpose of the project was to visualise what companies could do to broaden the base of the travel market in order to increase business. The research facilitated the delineation of a personality-based, psychographic typology. Tourists are classified based on personality traits along a continuum, with allocentrics on one end of the spectrum and psychocentrics on the other. Allocentrics, also referred to as 'venturers', are tourists who are comfortable making choices that involve some degree of variation, adventure or risk. In contrast, psychocentrics, also referred to as 'dependables', try to make safe, consistent, choices by preferring popular activities. Somewhere in the middle of the continuum are the mid-centrics, tourists who have a balanced combination of both psychographic traits. Tourists might lean towards one direction or the other on the continuum and are not dichotomously polarised.

According to Plog's model, destinations typically move through this continuum, appealing to the various tourist types as part of the process. Early tourists who effectively 'discover' a new place are likely to be allocentrics. They start talking to other people about their holidays, marketing by word-of-mouth, and recommending the exotic place they visited. This leads to a larger number of people traveling to the still under-developed area, which is then visited by near-allocentrics. After some time, as the destination becomes more popular, more tourist infrastructure is provided and marketing and promotion activities increase, eventually leading to the introduction of mid-centrics, which consequently stimulates further development of the area and the destination reaches a point in which it is widely popular with a well-established image that attracts mass tourism, thus appealing to the psychocentrics. This can be said for TEFL tourism, where some tourists will choose to be located in destinations that have a developed tourist infrastructure, such as Phuket,

Thailand, whereas others may choose to be based in remote, non-touristic destinations, facilitating a more unique cultural and adventurous experience.

While Cohen (1972) and Plog (1987) introduced the notion of the tourist typology, the following years saw the face of the industry change dramatically. Niche tourism forms were introduced followed by micro-niche tourism forms. This book has strongly associated TEFL tourism with volunteer tourism, owing to its similarities. Thus, the formation of the TEFL tourist typology is also underpinned by research undertaken within the area of volunteer tourism.

One area of research that was deemed particularly relevant to the study of TEFL tourists was surrounding the concept of *behavioural inversion*. Researchers have argued that tourists tend to behave differently when away from home. This phenomenon contributes to the suppression of personal limits, while favouring aspects such as the abuse of alcohol and/or drugs (de Oliveira and Paiva, 2007; Lomba *et al.*, 2009). Differences in behaviour may in part be due to individuals being far from home and therefore not being confined by the constraints of family and duties which normally moderate their behaviour (Redmon, 2003). Other contributing factors may be that alcohol and drugs purchased abroad may differ in strength from tourists' home countries and hotter climates enhance the risk of dehydration and disorientation (Bellis *et al.*, 2009). Some view this type of tourism as a form of risk-taking (Uriely and Belhassen, 2005), with tourists perceiving such holiday activities as a 'licence for thrill' (Wickens, 1997).

Malam (2004) states that identities change across space and context, supporting the notion that the role of a volunteer tourist shifts from seeking out pleasure, relaxation and stimulation during part of their trip to altruistically helping the community or environment at other times (Mustonen, 2005). For example, it can be argued that a tourist may not necessarily undertake their trip based on motives to take drugs; instead this may comprise only one element of their trip (Uriely and Belhassen, 2005). Furthermore, tourists may not necessarily have pre-planned

**Box 7.6.** Teaching in rural Thailand

For the 'allocentric' type of TEFL tourists, teaching in rural Thailand is perfect. There are many unexplored areas such as valleys, caves and winding roads to explore by moped. The locals may not speak any English – which is something that many tourists do not come across often nowadays. The food is authentic Thai (i.e. it is very spicy!).

These tourists have no full moon parties with glow paint and buckets, instead they party with the locals, having water fights during *Songkran* or releasing Chinese lanterns during *Loi Krathong*.

These tourists will inevitably learn the language, as difficult as this may be, because they have no option. They are entirely immersed into the local community and culture and may be without many of their home comforts.

These tourists are able to achieve an experience that is difficult to come by in today's society, where the world is increasingly globalised, particularly in areas which cater for tourists. Here there is likely no tourist provision and the TEFL tourist is viewed as the commodity, not the other way around.

**Box 7.7.** Cultural integration in TEFL tourism

TEFL tourism is unique in that it provides the tourist with a cultural experience not usually available through alternative tourist experiences.

It enables the tourist to spend time with the local community outside of the realms of the tourist industry, where staff aim to please and to satisfy the tourist's needs.

Working in the classroom enables an authentic experience which takes place on a level deep enough for the TEFL tourist to obtain a true and thorough understanding of the working of the community within which they are based.

Cultural areas such as food, music and language are used constantly around the TEFL teacher, allowing them to develop a 'rich' understanding unlike that which would be developed by visiting museums, historical buildings or temples alone, which are common tourist activities.

consumption of these substances as suggested by Valdez and Sifaneck (1997). Instead, Uriely and Belhassen (2005) claim that tourists who are attracted to a destination through awareness of drug accessibility and those who only become aware of this during their visit both classify as drug tourists.

It is important to understand that not all TEFL tourists are the same and to put them into a metaphorical box would be unjust. It does, however, make it easier in terms of appropriate management if tourists are typologised in a way that can enable easy identification of typical traits and behaviours. Despite criticisms of the use of typologies in tourism, these remain an integral and frequently cited element of the body of tourism literature. They also provide an indication of tourist types which can then be examined explicitly in specific contexts. The prominent use of typologies in both the tourism literature and in particular niches such as volunteer, cultural and educational tourism, support the development of the TEFL tourist typology in helping to achieve a broad understanding of the TEFL tourism industry and the tourists within it.

The model below (Fig. 7.3) demonstrates the typology developed for TEFL tourists operating in Thailand. While there are some specific elements, such as the fact that there is a substantial elderly, male population due to the nature of the expatriate community, many of these areas can be generalised across different contexts and geographical areas.

There were four dominant positions adopted by TEFL teachers in Thailand that centred around the concepts of leisure, philanthropy, career and expatriatism. The typological models were created through undertaking a series of four logistic

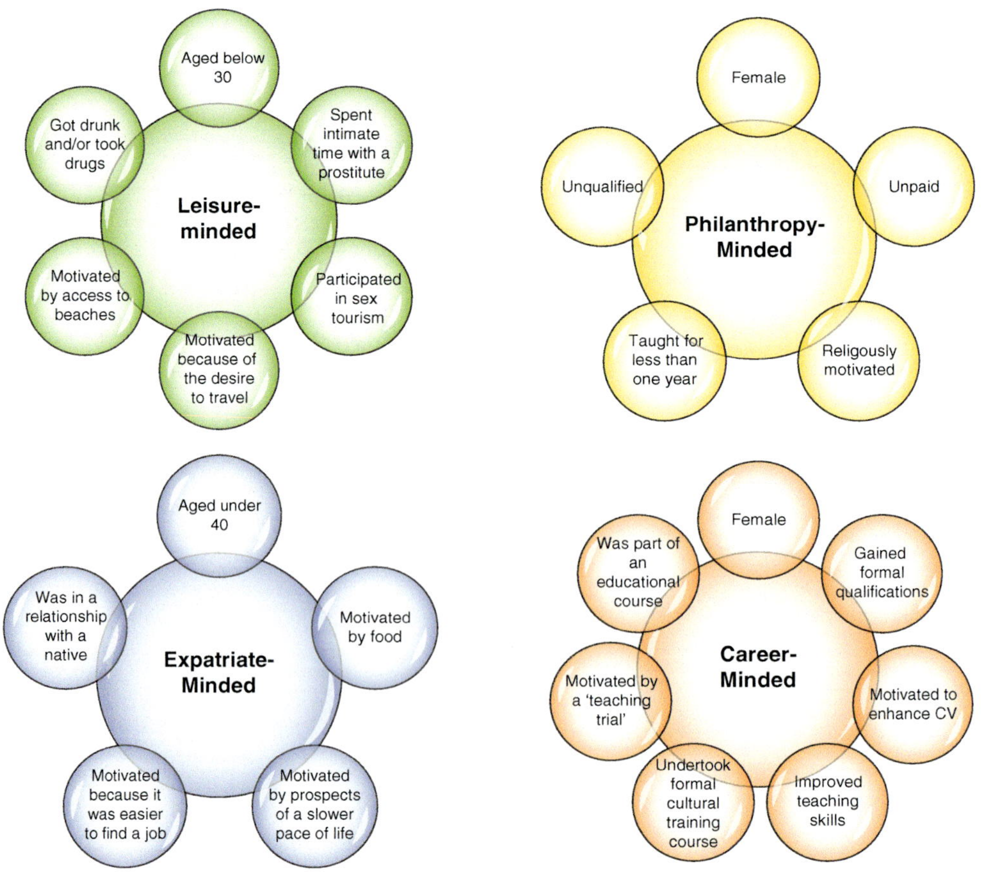


**Fig. 7.3.** TEFL tourist typology

regression tests which examined the strength of the influence that each variable had on the model (see Stainton, 2018b).

The majority of respondents demonstrated attributes associated with all positions throughout their TEFL experience, which shows that they do not neatly fit into any one category, although one may be more prominent than another. This supports the notion of role ambiguity (Lyons and Wearing, 2008), where it is unclear who the TEFL tourist is and how to define them based on the ability to possess multiple motivations simultaneously (Hustinx, 2001; Tomazos and Butler, 2010). Some scholars have suggested that rather than categorising tourists de facto, it is best to position them along a continuum (e.g. Hustinx, 2001; Tomazos and Butler, 2010). However, this would not account for the shifting roles (Mustonen, 2005; Sin, 2009) of TEFL teachers. It was found that TEFL teachers have varying roles depending on the day of the week, geographical location or the company they are in. For example, a TEFL teacher may be philanthropy- or career-minded Monday to Friday when in the local community, and leisure-minded at the weekend when they travel to the tourist areas.

This changing of roles demonstrates behavioural inversion (de Oliveira and Paiva, 2007), where bloggers and respondents displayed behaviour that they would not usually exhibit in their educational environment. Effectively, they appeared to adapt their identities depending upon space and context (Malam, 2004), and whether they had the constraints of teaching responsibilities confining them to moderate their behaviour accordingly (Redmon, 2003). For opportunities outside the community within which they teach TEFL tourists are able to be transported away from the restraints of their everyday life into states of abandon (Tutenges, 2013).

The first two teacher types, the leisure-minded and the philanthropy-minded TEFL tourist, largely correspond with existing volunteer tourism typologies that categorise tourists according to their motivation. Similarly to Brown and Morrison's (2003) vacation-minded and Daldeniz and Hampton's (2011) volunTOURISTs, the leisure-minded TEFL tourist tends to be travel-orientated. They are motivated by leisure opportunities including beach and nightlife tourism. They are likely to drink alcohol and take drugs as part of their TEFL experience and it is common that they have some involvement with the sex trade, ranging from casual attendance of sex shows to spending intimate time with a prostitute. This group often consists of tourists aged below 30.

**Box 7.8.** Role ambiguity in TEFL tourism

Felix was the model teacher. He taught across the entire spectrum at his secondary school in Ghana and his students were achieving very high grades at the end of each term. Felix was on time for all of his classes, dressed smartly and was highly regarded by both staff and students.

On the weekends, however, it was a totally different story. Felix enjoyed partying and frequented a local bar/nightclub in the village. He was known by the locals as 'the life of the party' and never missed a social event. He could drink any Ghanian under the table and he gave them a good run for their money with his weed smoking abilities too. From Friday evening through to Monday morning he spent most of his time in a cloud of marijuana fumes, intoxicated by drugs and alcohol.

This is the perfect example of behavioural inversion. Felix effectively had two identities: 'teacher Felix' and 'party Felix'.

**Box 7.9.** The TEFL 'bum'

Sadly, not all TEFL tourists are passionate about teaching. Some do it simply to fund or to extend their travels. In fact, there is a term sometimes used for these types of TEFL teachers: the TEFL 'bum'.

This is a person who is typically lazy and provides a sub-standard quality of teaching. They may spend their evenings in the local bar or at a party and their weekends chilling on the beach or smoking cannabis.

This TEFL tourist is not career-driven (not at this stage in their life, anyway) and they are not work-focused.

It is likely that this person has a long-term travel agenda, whether they live a nomadic lifestyle or whether they plan to stay in that particular location long term.

This is not the type of teacher who is likely to yield optimum educational outcomes from their students, but who students may favour as they are likely to take a relaxed approach to aspects such as homework (which involves additional work for him or her).

Likewise, the philanthropy-minded TEFL tourist corresponds with Brown and Morrison's (2003) volunteer-minded and Daldeniz and Hampton's (2011) VOLUNtourists. This teacher type is often religiously motivated. They may be on a short-term, unpaid placement and without the formal requirements usually needed to work as a TEFL teacher. They are often female. This is the smallest group of TEFL teachers, perhaps where philanthropy is not the primary motivation (Lyons and Wearing, 2012), or where the TEFL tourist puts on an altruistic facade to disguise their true motivations (Wearing and McGehee, 2013). There is also the concept of philanthro-capitalism (Hero, 2001; McAlister and Ferrell, 2002; Edwards, 2009), where philanthropy is used as a means to enhance one's competitive identity, thus indicating that although intentions may appear altruistic, they are in fact selfish.

In the Stainton study, approximately half of TEFL tourists were classified as career-minded, making this the largest group. This teacher type closely corresponds with Ritchie *et al.*'s (2003) description of educational tourism, as education and learning are important parts of the TEFL experience. The education sphere takes two approaches: the education *of* the TEFL teacher and the education *provided by* the TEFL teacher. The career-minded TEFL tourist centres around the former. This perhaps fits best with the education first tourist noted by Ritchie *et al.* (2003), although it is not necessarily limited to university, college or school tourism as they suggest. It also appears to be akin to existential (Kolb, 1984), transformative (Mezirow, 1991; Coghlan and Gooch, 2011) or lifelong learning (Broomhall *et al.*, 2010; Falk *et al.*, 2012) that may be facilitated through a formal educational institution or may take place through a process of experience and reflection (Chapter Three discusses educational tourism theory). The notion of learning among TEFL teachers as opposed to their students gives rise to the question as to whether the respondents are also students in their own regard, placing the concept of education in a new light within the TEFL experience. Most career-minded TEFL tourists undertake formal study and qualifications as part of their experience, demonstrate improvement in their teaching skills and are motivated by the prospects of a trial teaching period before potentially commencing a teaching career in their home country or elsewhere. CV enhancement is also a factor. They are often female.

The expatriate-minded TEFL tourist in the case of Thailand is mainly male. The typological model demonstrated that expatriate-minded TEFL tourists tend to be motivated by lifestyle prospects including food, the cost of living, job opportunities and the prospects of a slower pace of life than they would otherwise have in their home countries. The expatriate-minded TEFL tourist is likely to differ according to location and typical expatriate communities.

# 8

# The Future of the Industry

Throughout this book, two fundamental points have been emphasised. Firstly, TEFL is big business. Global demand for English language abilities is continuing to grow as English language has become an integral part of world society. Governments are now increasingly recognising the importance of English to their economies and societies, and individuals often see English as a tool that can help them to fulfil their personal aspirations and ambitions. In fact, English has become a centrepiece of educational reform in many countries, often in response to developments in economics, demographics and technology. Improving a nation's English literacy is not only important in terms of business, education and diplomacy, but also to enable the populace to enjoy culture and entertainment.

From a business perspective, being able to communicate in English is an increasingly important skill as world markets become more interconnected than ever before. As new economies develop rapidly, the value of being able to speak Mandarin, Arabic and Hindi is increasing but, for now, English remains the leading international language. The dominance of English is evident throughout many aspects of life including scholarship, the internet and technology, the global media and advertising, international politics and diplomacy and, particularly, global business. The evidence for the growth of English-language speaking and the demand for teachers is indisputable.

Secondly, this book has introduced TEFL teaching within the context of tourism, a concept surprisingly unrecognised within either the educational or the tourism industries. It has explored this notion through identifying a number of similarities between TEFL tourism and various tourism forms, placing it firmly in the 'micro-niche' category. Links were identified with a number of macro-niche tourism forms including volunteer, philanthropic, cultural, pro-poor, educational, adventure and gap year or backpacker travel. It is also clear that within the concept of TEFL tourism, notions of teaching, linguistics and service learning make up inherent parts of the experience.

This book has assessed many aspects of TEFL tourism, raising concerns where appropriate. Drawing on experiences from the wider tourism industry and the education sector, a number of sustainability issues have been highlighted. However, this text just touches the surfaces of the topic, and there is indeed an urgent requirement for TEFL stakeholders to review their operations and practices in order to ensure that sustainable and optimum management is undertaken, where opportunities and positive impacts are capitalised upon and negative impacts are minimalised. In this light, this final chapter assesses the future possibilities for TEFL tourism, highlights lessons learned from the wider tourism industry and suggests the implementation of appropriate guidelines, regulations and mindfulness.

## Future Growth of the TEFL Tourism Industry

In 2014, the British Council estimated that the number of English language learners worldwide will increase to over 2 billion by 2020. All countries are integrally locked into the global economy and no country can succeed without operating in all the major established and emerging markets, thus fuelling the need for global communications; this is only set to increase in future years. According to the International TEFL Academy (2018), 88% of English teachers abroad in non-native English speaking countries are not native English speakers themselves because there are simply not enough native English speaking teachers to meet demand. This book has demonstrated that this is the primary reason for the introduction of the concept of TEFL tourism – but is it enough?

With the continued demand for TEFL teachers overseas, it is likely that some destinations, particularly those which are less attractive to tourists for reasons such as unstable political situations, the presence of diseases such as malaria, undesirable weather, low wages or a lack of tourist infrastructure, will struggle to fill their vacancies. This is likely to have one of two major consequences. Firstly, the destination fails to develop economically and culturally at the rate of other nations in a similar position. Children in that destination are not presented with the opportunities that they may have had, had they been born in a different village, city or country. While this has always been an issue, the gap between the rich and the poor and between the developed and developing continues to widen.

The second outcome is that teaching positions are filled by unsuitable candidates. Teachers may be employed simply because they are natives of an English-speaking country, essentially 'qualifying' them to be appropriate for the position. However, they may lack the skills and experience to undertake the required workload effectively. Efficiency may be reduced because of poor qualities of staff and learning outcomes. Some TEFL tourists may come and go quickly, resulting in high turnover. Not only does staff recruitment cost the educational organisation money, but it also disrupts the education of the learners involved. Overall, hiring ineffective teachers may not yield the positive educational outcomes desired and in some instances may even impede on education that would have taken place instead of the English session, should this not form part of the curriculum.

This is not to proclaim that this is the future of TEFL tourism, but to emphasise that sustainable management is very important. Sustainable management is effectively the intersection of business and sustainability. It requires assessment of the long-term viability of the industry and its practices together with analysis of any associated impacts. The intention for sustainable management is to be preventative, rather than reactive, although this is sadly not the case with many tourism businesses where negative impacts are evidenced all too frequently.

Traditionally, business leaders concern themselves most with their bottom lines, or in other words, the monetary profits of their business. Nowadays, sustainability is a key consideration driven by a number of forces including government regulations or incentives, corporate social responsibility and a general sense of increased awareness. The triple bottom line theory expands the traditional accounting framework to include two additional performance areas: social and environmental impacts, shown in Fig. 8.1.

The concept of people plays a particularly important role within TEFL tourism, as people are fundamental to the industry. The triple bottom line theory refers to people in a range of contexts including employees or TEFL tourists, members of the host community, and other relevant stakeholders. If a company adopts a triple bottom line way of working then it should offer fair wages to TEFL tourists and other employees, be ethical in its marketing (i.e. inform prospective tourists of the experience and what it is likely to

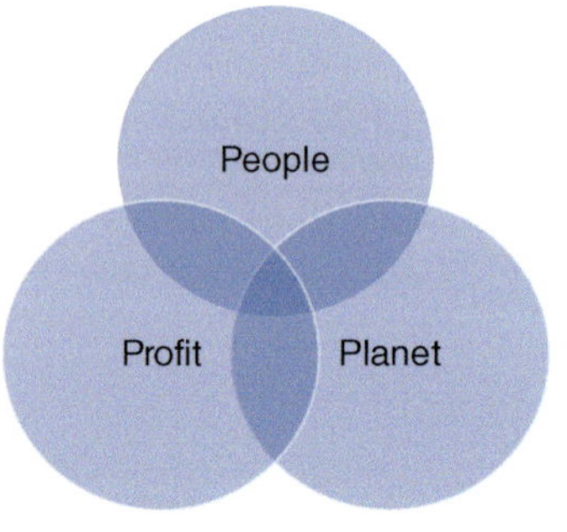


**Fig. 8.1.** The triple bottom line

entail, as discussed earlier in this book), ensure fair working conditions for teachers, and that outputs are appropriate and beneficial for the local community in which it serves.

The planet is not an area that holds much weight within TEFL tourism, given that it does not have a significant impact on the environment. There are, of course, some elements worthy of consideration that would apply to all businesses, such as reducing waste, investing in renewable energy, managing natural resources more efficiently and improving logistics that TEFL organisations and stakeholders should take into consideration in order to be sustainable.

While every business pursues financial profitability, triple bottom line businesses see it as one aspect of a business plan. Sustainable businesses recognise that profit is not diametrically opposed to people or planet and that by prioritising all three aspects collectively, better outcomes will be achieved. The problem with TEFL tourism lies in the differences in priorities between the public sector and for-profit organisations.

Given the number of stakeholders involved in TEFL tourism, it is fundamental that relationships are developed and that they operate effectively. A public-private partnership (PPP) is a cooperative arrangement between two or more public and private sectors, typically of a long-term nature. While there is no conclusive definition of PPP, it can be understood both as a governance mechanism and a brand strategy. PPP can cover a vast array of different types of contract, which are frequently of long-term duration, with a wide range of risk allocations, funding arrangements and transparency requirements. As a brand, the PPP concept is closely related to aspects of privatisation and the contracting out of government services. From a governance perspective, it encompasses a number of potential arrangements, typically associated with identified community or national needs.

PPPs are beneficial in that they allow for risk-sharing and the development of innovative, long-term relationships between the public and private sectors. One of the key motivations for government-based staff to embark on private–public relationships is that financial capital can be afforded to them. While PPP has been controversial, it is clear that in the area of TEFL tourism, it allows a government to fill the employment gap without conjoining with the private sector.

In idealistic terms, PPP in TEFL tourism will allow for government strategies to enhance English language capabilities and the associated benefits while providing the opportunity for foreign investment in TEFL agencies and stakeholders. The reality is, however, that the private enterprises facilitating TEFL tourism may be profit-orientated, may not focus their attentions on the specific needs of the community and may not recruit and send the highest quality TEFL teachers. This is a result of the contradicting nature of private and public operations.

Alongside PPP, there is a need for all TEFL tourism stakeholders to work carefully together. Stakeholder management is the systematic identification, analysis, planning and implementation of actions designed to engage with stakeholders and is arguably the most important ingredient for successful business. There are several key principles that are important to consider.

1. Stakeholders should have a say in decisions that affect them (seeks out those potentially affected by, or interested in, a decision).
2. Stakeholder participation includes the promise that their contributions will influence decisions (seeks input on how they may wish to participate).
3. Stakeholder engagement (provides information, time and space to allow those involved to participate in a meaningful way).

There are a variety of stakeholders involved in TEFL tourism including, but not limited to, those identified in Fig. 8.2. Each plays an integral role in contributing to the industry's success and sustainability. To summarise, this includes the following:

- ensuring that teacher training providers and universities in tourist-sending destinations are aware of what the experience entails and which elements of their curriculum may or may not be encompassed in the TEFL tourism experience;
- ensuring that TEFL tourists are appropriately qualified to undertake the role of teacher, have the correct attitudes and are not going to promote or enhance any negative impacts associated with TEFL tourism;

- maintaining positive relations with the host community, whether directly or indirectly involved in TEFL tourism; and
- ensuring TEFL agencies take into consideration the people, planet and profit elements of their business.

As demonstrated throughout this book, the TEFL tourist is perhaps the key stakeholder and thus it is important to recognise the changing nature of tourist wants and desires in the future. As identified by Yeoman (2012), tomorrow's tourists lead complicated lives and have a fluid identity. They seek new experiences, luxury, culture and authenticity in ways that they have not previously. The tourist wants to sample the ethnicity of the destination; increasingly, interests in culture, food and sport are shaping the way people approach their travel choices. Tourists seek new challenges and a heterogeneous product. TEFL tourism satisfies these future demands in that it is unique and polarised in many ways from the traditional sense of 'holiday'. However, as more and more micro-niche tourism forms develop, it will need to ensure, as an industry, that it remains current in order to satisfy changing needs and demands.

The last point that warrants attention in the discussion of the future of TEFL tourism is the changing dynamics of the TEFL community. The TEFL sector is in the preliminary stages of even further disruption, where tourists are not only filling demands in the classroom, but are also taking their expertise (or perhaps lack of it) online. The growth of the internet and technological advances around the world has facilitated the rise of a new market for English language learning. Opportunities for online TEFL teachers have increased year on year as more and more students have turned to learning online. While school-aged students are generally supplementing the traditional school-based learning approach with educational apps and other digital products, more and more adults are moving to online classes. With geographical

TEFL Stakeholders

**Education overseas**

- Universities
- TEFL providers
- Educational institutions/schools
- Future employers

**Education In-destination**

- National, regional and local government
- Local community
- Suppliers

**TEFL businesses**

- Business owner(s)
- Employees
- Contractors

TEFL Tourist

**Fig. 8.2.** TEFL stakeholders

**Box 8.1.** TEFL apps in developing countries

Socrative is a cloud-based student response system. It was developed in 2010 by graduates based in Boston. It enables teachers to create simple quizzes that students can take quickly on laptops, tablets or their own smartphones.

Quizzes can feature questions of various natures such as true/false, multiple choice, graded short answer or open-ended short responses. Activities can be teacher-paced or student-paced. There's also a gaming element enabling teams of students to compete against one another. Socrative quizzes can also be shared with other teachers and there is a library of 315,000+ quizzes that any teacher can use.

Socrative officials say more than 350,000 teachers have registered for free accounts, and that more than 1700 districts or schools count at least 20 teachers as users. Those teachers are pretty active: there [were] over 6.2 million student logins in 2012.

(Edsurge, 2018)

limitations no longer a factor, ESL teachers can now work from anywhere in the world.

The convenience of online TEFL teaching is surely a big draw for both students and teachers. As the technology around online tutoring improves and as parents and the public become better informed about online options to learn English, the demand for online tuition for English learners will certainly increase. This further changes the face of the TEFL tourism industry, where teachers need not reside in the location in which they are teaching. This could be beneficial in destinations which are not viewed as being particularly desirable to TEFL tourists, although it could also enhance levels of standardisation which disallows for differentiation accorded to culture and location and which encourages standardised pricing regardless of the cost of living in certain localities, rendering it difficult for said learners to participate.

## Lessons Learned from the Niche Tourism Umbrella

Niche tourism has traditionally been hailed as being a more sustainable approach to tourism development than its mass tourism counterpart; however this does not mean that it is without its pitfalls. As previously noted, TEFL tourism has close ties with volunteer tourism, the main difference being the presence of financial remuneration or incentives. In fact, referring to the body of literature addressing the volunteer tourism industry is particularly useful in beginning to attempt to comprehend the TEFL tourism industry over the past 20 years. Increasing scrutiny and critical assessment of volunteer agencies and operations has brought to light a number of areas warranting careful attention and, given the similarities between the TEFL and volunteer tourism sectors, there are many lessons that can be learned from this macro-niche and applied to the micro-niche of TEFL tourism from both the perspective of the TEFL tourist and the local community.

Although TEFL tourists may benefit through feeling that they have 'made a difference', Butcher (2011) highlights in the context of volunteer tourism that believing that this is the case is in fact naïve, self-centred and arrogant, as it is often not representative of reality. For example, some volunteers are in search of an authentic experience (Sin, 2009; Wickens, 2011); however if their experience is in fact inauthentic or staged in some way, it creates ethical concerns for the volunteer. In severe cases, the need for volunteers may even be simulated (Tomazos and Cooper, 2012), and this has been exemplified through the recent media stories surrounding volunteering in orphanages, where it has been identified that many children are in fact taken away from their families and 'orphaned' in order to serve the volunteer tourism industry and to make a profit (Tourism Concern, 2014). Such examples provoke questions as to whether it is moral to allow the volunteer to believe they are helping a destination or community, when in reality they may be making little difference, or in extreme cases even worsening local development (Benson and Wearing, 2012). It also highlights the need for further regulation of the sector and for additional research into the associated authenticity issues.

A limited number of studies have identified that volunteers have expressed dissatisfaction with their placements, largely due to their expectations not being met (Gray and Campbell, 2009; Guttentag, 2009; Coren and Gray, 2012). This is evidenced through Coren and Gray's (2012) identification that volunteers were aware that the money they had paid for their placement had not reached the Thai and Vietnamese communities that they had been led to believe it would. Similarly, volunteers in Palacios' (2010) research felt frustrated because they did not feel that they had the skills required to do the job. This is in contrast to the expected benefit of being able to do jobs they would not normally be able to do at home (Soderman and Snead, 2008).

There is also some reference to TEFL teaching surrounding this issue. Firstly, Coren and Gray's (2012) study of volunteers in Thailand demonstrated that although at face value they believed they would gain invaluable experience needed to be a teacher, at the end of the trip they did not feel that this expectation had been fulfilled. Meanwhile, TEFL teachers in Palacios' (2010) research found that they were treated as 'experts', with local teachers asking for their help and resources. This subjected them to unequal power relations and made them feel uncomfortable. Thirdly, Mostafanezhad (2013) noted the way in which the differences in the education systems were often a culture shock to volunteer teachers. These three examples highlight that the TEFL teaching sector may not be as beneficial to the participant as initially perceived and, owing to the limited studies, there is an urgent call for further research and clarification in this field.

From the point of view of the host community, one of the main issues arising in the literature is employment. Some studies find that locals are displaced by volunteers (Benson and Wearing, 2012) or have experienced employment losses (Ellis, 2003). Volunteer tourists often perform jobs that locals could do (Guttentag, 2009), making the role of the volunteer unclear to the hosts (Palacios, 2010). As a result of volunteers undertaking various work in the community, as opposed to the local population, dependency can also be promoted (McGehee and Andereck, 2008; Guttentag, 2009).

The issue of employment is also addressed through the lack of skills required by volunteers. Although there are instances when volunteers do possess specific relevant skills, such as the volunteer psychologists in Palacios' (2010) study, this does not appear to be commonplace. Many volunteer tourism projects have minimal or no requirements regarding the skills set needed to participate (Brown and Morrison, 2003; Guttentag, 2009), and this can negatively impact on the quality of work (Guttentag, 2009; Benson and Wearing, 2012), and thus undermine the positive impacts desired.

The consequences of volunteers possessing limited skills do not appear to have been sufficiently examined within the current literature and there appears to be no reference to the TEFL teaching sector whatsoever. Naturally, the value of student education provided by an unqualified and inexperienced teacher is questionable, making it ironic that this does not appear to be a concern, nor a research topic warranting interest within the literature to date. Therefore, there is an urgent need for further research into this area.

Examination of Richter and Norman's (2010) study identifying the negative impacts associated with volunteering with children is also relevant to the TEFL teaching sector. They found that the formation of an intense bond between a volunteer tourist and a child can be devastating when the tourist leaves. Despite their research focusing on AIDS orphans, there will also be an inevitable bond established between volunteer TEFL teachers and students. As many volunteer tourism placements are short-term, this potential negative impact is another area that warrants further research attention.

It is often assumed that communities will favour volunteer tourism projects and their intended benefits (Guttentag, 2009), but this may not always be the case. For example, Matthews' (2008) study of turtle conservation in Costa Rica identified that the volunteer tourism project had caused poachers and subsequent stallholders selling turtle products to lose their livelihoods. Of course, poaching is not commendable, however this example does highlight that while the benefits of volunteer tourism (in this case conservation of turtles) must be considered, it is important that any volunteer tourism projects are designed in line with the host community's wishes and needs.

Volunteer projects are more likely to succeed when active local level support exists (Benson and Wearing, 2012; Wearing and McGehee, 2013), and when careful planning has taken place in order to minimise the possible negative impacts. It is imperative to recognise and address such possibilities when planning for volunteer tourism in order to promote sustainability of the industry.

Wearing (2001) argues that contribution to development cannot be measured simply in terms of the projects themselves. Rather, volunteer tourism helps to develop the people who will potentially act ethically in favour of those less well-off during the course of their careers and lives. Correlating with the prior argument suggesting that the benefits of volunteer tourism permeate beyond the trip itself, this highlights the potential for a ripple effect of the intangible benefits of volunteer tourism not only for the tourist, but also for the host and others involved in the volunteer's life.

Measuring 'development' is a difficult task and the value of projects and the good they are doing is not always clear (Tomazos and Butler, 2009). It has been suggested that the development agendas of many organisations are in fact unrealistic and the language of 'making a difference' is simply tying the programmes to unachievable development goals. In contrast, it can be questioned whether the volunteer tourism industry would indeed survive without reference to such potentially unrealistic expectations (Palacios, 2010).

It is important to recognise that the needs and desires of host communities are varied, and as such it is naïve and counterproductive to generalise the needs and success of volunteer tourism (Guttentag, 2009), as the literature to date appears to have done to a large extent. This indicates the requirment for a more segmented approach to the different volunteer tourism research areas, highlighted by Stainton (2016). As shown, there are many negative impacts associated with volunteer tourism, as there are with the travel and tourism industry in general and with the varying other niche forms. While lessons can be learned from these allied industries, it is imperative that stakeholders and managers take into consideration the negative consequences resulting from the actions of tourism enterprises and individuals, and learn from these lessons so as to enable the TEFL tourism industry to be proactive rather than reactive.

## Guidelines, Regulations and Mindfulness

It would be naïve to believe that all TEFL agencies and stakeholders will prioritise people, planet and profit in the way that is suggested in this chapter. This is not only as a result of ignorance or selfishness, but also due to a lack of awareness. Anybody could start up a TEFL business, regardless of their qualifications or experience, and they may be completely unaware of the sustainability concerns discussed throughout this book. In an ideal world, the answer would be to regulate the industry, ensuring standardisation across teaching quality, recruitment processes and TEFL qualifications themselves. To achieve this across the globe is, however, an almost impossible endeavour.

Currently, the TEFL industry is succinctly unregulated. There is little clarity over the job description and person specification of a TEFL teacher, nor are there any standardised processes regarding contractual agreements, pay scales, rights or working conditions. It is aspects such as these that are the precursor to the grey dynamics currently presented within the TEFL industry on a global scale. While to a certain extent this can be deemed inevitable due to the global complexities noted in the previous section, it can be extremely problematic for stakeholders involved with TEFL practices and the TEFL community. A lack of regulation can significantly and inversely contribute to the sustainable management of the TEFL tourism industry and thus it is imperative to bring said issues to the front of the English language teaching discussion.

A common problem in many parts of the world is the lack of an institutional framework for the development of TEFL tourism. Besides that, in many countries the presence of several government ministries which have intersecting responsibilities regarding the developmental, planning and marketing issues relating to recruiting TEFL teachers and managing TEFL tourism can make things difficult in terms of conflicting agendas. On the one hand, there is the need to balance the development of tourism and manage the attendant economic and social impacts and, on the other hand, there is the need to meet development agendas and manage the education sector in the best way possible to achieve this end.

The need for umbrella mechanisms that allow the relevant bodies to work successfully together in order to create the balance necessary for the development of TEFL tourism should be stressed and it is imperative that public, private and NGOs cooperate in order to achieve any kind of agreed regulation within the TEFL industry. This is, most certainly, however, easier said than done. The TEFL tourism industry is in its infancy and as such it can be argued that the organisations involved have not yet reached the desired levels of consciousness in order to maintain a sustainable and appropriately regulated industry.

The volunteer tourism industry has recognised this and, while regulation may be difficult to achieve, the Association of British Travel Agents (ABTA) has implemented some volunteer tourism guidelines for operators. These were developed in consultation with a wide range of organisations including international travel industry and membership organisations, NGOs, charities, leading academics, intergovernmental organisations and tour operators working in the volunteer tourism sector. The guidelines aim to tackle bad practice and improve volunteer tourism standards throughout the travel industry (Box 8.2).

**Box 8.2.** Organisations which have subscribed to ABTA's volunteer tourism guidelines

- Blue Venture Expeditions
- Born Free Foundation
- Bournemouth University
- Comhlámh
- ECPAT UK
- Hope and Homes for Children
- International Centre for Responsible Tourism
- Real Gap
- Kasbah du Toubkal
- Kuoni Travel
- Leeds Beckett University
- Madventurer
- Moving Mountains
- Oxford Brookes University
- People and Places
- POD Volunteer
- Responsible Travel
- STA Travel
- The Seaver Foundation
- The Travel Foundation
- University of Brighton
- University of Surrey
- UNWTO
- World Challenge
- Year Out Group
- This Life Cambodia
- Global Alliance for Children – Cambodia
- M'lup Russey
- The Code
- Better Volunteering, Better Care
- Friends International – Laos/Cambodia
- NGN Nepal
- Projects Abroad
- UNICEF
- Friends International
- Planeterra Foundation/G Adventures Base Camp
- Kaya Volunteering
- Do Good as You Go/The Muskoka Foundation
- Sue Berelowitz: Visiting Professor, Institute of Applied Social Research, University of Bedfordshire
- Philip Mudge: Comhlámh
- Rachel McCaffery
- Anna McKeon: Better Volunteering, Better Care

ABTA's volunteer tourism guidelines build on the wide range of existing guidance available throughout the tourism industry. The aim is to encourage businesses to achieve high standards in the volunteering experiences they offer, which will result in improved impacts for stakeholders. The guidelines cover a wide range of issues, including putting adequate safeguards in place to protect the rights of children and vulnerable people, and checking there are similar safeguards to protect animals and wildlife. In order to capitalise on the possibilities for sustainable development in TEFL tourism, it is recommended that a set of guidelines is proposed and developed similar to the volunteer tourism industry. In this way, TEFL tourism stakeholders, including the government, TEFL agencies and indeed the education industry would have some good practice principles to aspire to achieve. By using guidelines in this way, stakeholders would also have the potential to contextualise these according to the geographic location, cultural context and educational priorities in the destination. While this will not solve all of the potential problems associated with TEFL tourism, it would be a good start along the path.

## Concluding Remarks

In summary, this book has aimed to provide insight and clarity about the industry to all involved in TEFL tourism. It is clear that the industry has a strong future, although the shape and face of this may change over the years, especially given the growth of the online marketplace. The scale of the industry is indisputable, as is the need to ensure sustainable management. There is a wealth of evidence to demonstrate the pitfalls of poor management in both the education sector and within the tourism industry. Therefore, this book has introduced a range of concepts and highlighted a number of key concerns worthy of additional consideration.

In a world of increased globalisation and interconnectivity, the power of the English language is stronger than ever before. TEFL tourism is a technique for meeting the demands of English language learning that has been used for a number of years, yet it has to date not been recognised as a tourism form. Many niche forms of tourism, such as the volunteer tourism industry, have developed in such a way that they can now be classified as macro-niches, leaving room for emerging areas of tourism, such as TEFL tourism, to fill their metaphorical shoes.

The tourism industry is arguably one of the largest in the world and there is a wealth of experience that TEFL stakeholders can learn from in order to capitalise on the development and management of this growing micro-niche tourism form. While this book has focused strongly on the concept of comparison with volunteer tourism, this is not the only area of tourism that the industry can draw upon in learning how best to manage this growing niche.

English language learning has in many ways become commodified and now presents similarities with the notions of package tourism. The number of TEFL agencies is growing exponentially and, in many cases, they are somewhat replicated by the typical tour operator which compiles and sells package holidays. There is a distinct paradox, however, between the benevolent intentions commonly held by the not-for-profit sector and the profit-orientated focuses of the private sector. This is something that requires careful management in order to obtain the optimum outcomes for all involved. This is best achieved through careful examination of associated tourism forms, the education sector in the destination under scrutiny, the TEFL operators involved and the TEFL teachers themselves.

Perhaps the biggest concern within the TEFL tourism industry is the amount of ambiguity. There is no clear role of a TEFL teacher and TEFL teaching varies considerably according to the educational and cultural context in which it is situated; there are no regulations and no clear guidelines to work to. There are many risks at play here, such as hiring TEFL teachers who in their home country would not be cleared to work with children due to a criminal record or people who have simply purchased TEFL qualification certificates. This book has highlighted many areas such as this that require clarity and uniform procedures or guidelines.

In summary, therefore, this book calls to industry for stakeholders to work together to ensure a sustainable future for TEFL tourism. It has presented a number of case study and practical examples to help make comprehension of the industry accessible both to academics and industry practitioners. Additional research is vital, including further examination of the concept of the 'TEFL package' and the commodification of the product, the possible negative impacts for the host community, the possibility of implementing some regulation or examination of motives and experiences of TEFL tourists themselves. The book has aimed to serve the academic communities of both the education and tourism spheres as well as introducing the concept of TEFL tourism to industry professionals and the wider public.

# References

Abrams, I. (1979) The Impact of Antioch Education through Experience Abroad. *Alternative Higher Education* 3, 176–187.
ACCET (2018) Available at: https://accet.org/about-us (accessed 11 March 2019).
Adams, J.S. (1963) Towards an Understanding of Inequality. *Journal of Abnormal and Normal Social Psychology* 67, 422–436.
Adventure Travel Trade Association. (2018) *Adventure Travel*. Available at: https://www.adventuretravel.biz/ (accessed 15 January 2018).
Alderfer, C. (1969) An Empirical Test of a New Theory of Human Needs. *Organizational Behavior & Human Performance* 4, 142–175.
Alexander, Z. (2012) The Impact of a Volunteer Tourism Experience, in South Africa, on the Tourist: The Influence of Age, Gender, Project Type and Length of Stay. *Tourism Management Perspectives* 4, 119–126.
Allender, J. and Allender, D. (2006) How Did Our Early Education Determine Who We Are As Teachers? In: Fitzgerald, L.M. Heston, M.L. and Tidwell, D.L. (eds) *Collaboration and Community: Pushing Boundaries through Self-study,* Sixth International Conference on Self Study of Teacher Education Practices, Herstmonceux Castle, East Sussex, England, 14–17.
Appleby, R., Copley, K., Sithirajvongsa, S. and Pennycook, A. (2002) Language in Development Constrained: Three Contexts, *TESOL Quarterly* 36, 323–346.
Asia Teaching Jobs (2018) Available at: http://www.asiateachingjobs.com/jobs/tefl-trainer-for-8-days-at-tefl-heaven-koh-chang-1375674226-5945 (accessed 28 November 2018).
Balaam, D. and Dillman, B. (2013) *Introduction to International Political Economy*. London: Routledge.
Barrett Value Centre (2017) *The Barrett Model*. Available at: https://www.valuescentre.com/mapping-values/barrett-model (accessed 5 October 2018).
Basu, K., Maddox, B. and Robinson-Pant, A. (2009) *Interdisciplinary approaches to literacy and development*. London: Routledge.
BBC (2017) Available at: http://news.bbc.co.uk/1/hi/business/2828391.stm (accessed 2 June 2017).
Bellis, M., Hughes, K., Calafat, A., Juan, M., and Schnitzer, S. (2009) Relative Contributions of Holiday Location and Nationality to Changes in Recreational Drug Taking Behaviour: A Natural Experiment in the Balearic Islands. *European Addiction Research* 15, 78–86.
Bello, D. and Etzel, M. (1985) The Role of Novelty in the Pleasure Travel Experience. *Journal of Travel Research* 24, 20–26.
Benson, A. and Seibert, N. (2009) Volunteer Tourism: Motivations of German Participants in South Africa. *Annals of Leisure Research* 12 (3 and 4), 295–314.
Benson, A. and Wearing, S. (2012) Volunteer Tourism: Commodified Trend or New Phenomenon? In: Moufakkir, O. and Burns, P.M. (eds) *Controversies in Tourism*. Wallingford: CAB International.

Bentley, T. and Page, S. (2001) Scoping the Extent of Adventure Tourism Accidents. *Annals of Tourism Research* 28, 705–726.

Bergman, E. (2013) Philanthropy for Tourism and Development in Africa. *Synergos*. Available at: http://www.synergos.org/globalgivingmatters/features/1301philanthropyfortourism.htm (accessed 3 March 2016).

Bersales, J. (2005) Contested Space: Tourism, Power and Social Relations. In: Alampay, R. (ed.) *Sustainable Tourism, Challenges for the Philippines*. Phillipines: APEC.

Birdwell, J. (2011). Service International. London: Demos.

Bonink, C. (1992) Cultural Tourism Development and Government Policy. MA dissertation, Rijksuniversiteit, Utrecht.

Boorstin, O. (1964) *The Image: A Guide to Pseudo-Events in America*. New York: Harper & Row.

Borg, S. (1997) *Unifying Concepts in the Study of Teachers' Cognitive Structures*. Unpublished manuscript.

Borg, S. (2003) Teacher Cognition in Language Teaching: A Review of Research on what Language Teachers Think, Know, Believe and Do. *Language Teaching* 36, 81–109.

Boydell, T. (1976) *Experiential Learning*. Manchester: Department of Adult Education, University of Manchester.

British Council (2014) Thailand English Teaching Programme. http://www.britishcouncil.org/study-work-create/work-volunteer/thailand-english-teaching-project (accessed 9 May 2014).

British Council (2018a) Available at: https://www.britishcouncil.org/voices-magazine/eltons-english-language-professionals-building-buess (accessed 29 November 2018).

British Council (2018b) Available at: https://www.britishcouncil.org/sites/default/files/english-effect-report-v2.pdf (accessed 28 November 2018).

British Study Centres (2018) Available at: https://www.british-study.com/franchises/ (accessed 29 November 2018).

Britton, S. (1989) Tourism, Dependency and Development: A Mode of Analysis. In: Singh, T. Theuns, H. and Go, F. (eds) *Towards Appropriate Tourism: The Case of Developing Countries*, 93–116. Frankfurt: Peter Long.

Broad, S. and Jenkins, J. (2008) Gibbons in their Midst? Conservation Volunteers' Motivations at the Gibbon Rehabilitation Project, Phuket, Thailand. In: Lyons, K. and Wearing, S. (eds) *Journeys of Discovery in Volunteer Tourism: International Case Study Perspectives*. Cambridge: CAB International.

Broomhall, S., Pitman, T., Majocha, E. and McEwan, J. (2010) *Articulating Lifelong Learning in Tourism: Dialogue Between Humanities Scholars and Travel Providers*. Canberra: Australian Learning and Teaching Council.

Brown, G.T.L. (2004) Teachers' Conceptions of Assessment: Implications for Policy and Professional Development. *Assessment in Education* 11 (3), 301–318.

Brown, O. (1998) In Search of an Appropriate Form of Tourism for Africa: Lessons from the Past and Suggestions for the Future. *Tourism Management* 19, 237–245.

Brown, S. (2005) Travelling with a Purpose: Understanding the Motives and Benefits of Volunteer Vacationers. *Current Issues in Tourism* 8, 479–496.

Brown, S. and Morrison, A. (2003) Expanding Volunteer Vacation Participation – An Exploratory Study on the Mini-Mission Concept. *Tourism Recreation Research* 28, 73–82.

Brudney, J. (2000) The Effective Use of Volunteers: Best Practices for the Public Sector. *Law and Contemporary Problems* 62, 219–255.

Bruthiaux, P. (2002) Hold Your Courses: Language Education, Language Choice, and Economic Development. *TESOL Quarterly* 36, 275–296.

Buckley, J. (2012) Rush as a Key Motivation in Skilled Adventure Tourism: Resolving the Risk Recreation Paradox. *Tourism Management* 33, 961–970.

Buckley, R. (2007) Adventure Tourism Products: Price, Duration, Size, Skill, Remoteness. *Tourism Management* 28, 1428–1433.

Budeanu, A. (2005) Impacts and Responsibilities for Sustainable Tourism: A Tour Operator's Perspective. *Journal of Cleaner Production* 13, 89–97.

Buhalis, D. and Costa, C. (2006) *Tourism Business Frontiers, Consumers, Products and Industry*. Amsterdam: Elsevier Butterworth-Heinemann.

BUNAC (2018) Top 5 Reasons to do a TEFL Course and Teach English Abroad. Available at: https://www.bunac.org/uk/blog/bunac-blog/volunteering/top-5-reasons-to-do-a-tefl-course-and-teach-english-abroad (accessed 11 September 2018).

Butcher, J. (2011) Volunteer Tourism May not be as Good as it Seems. *Tourism Recreation Research* 36, 75–76.

Butler, R. (1980) The Concept of a Tourist Area Cycle of Evolution: Implications for Management of Resources. *Canadian Geographer* 24, 5–12.

Bywater, M. (1993) The Market for Cultural Tourism in Europe. *Travel and Tourism Analyst* 6, 30–46.

Calafat, A., Blay, N., Bellis, M., Hughes, K., Kokkevi, A., *et al.* (2010) *Tourism, Nightlife and Violence: A Cross Cultural Analysis and Preventative Recommendations*. Valencia: Martin Impresores.

Callanan, C. and Thomas, S. (2005) Volunteer Tourism: Deconstructing Volunteer Activities within a Dynamic Environment. In: M. Novelli (ed.) *Niche Tourism: Contemporary Issues, Trends and Cases*. Oxford: Butterworth-Heinemann.

Cambridge English Assessment (2018) Available at: https://www.cambridgeenglish.org/teaching-english/teaching-qualifications/celta/ (accessed 28 November 2018).

Carbone, G. (2004). *Tour Operators' Responsibility and Best Practices Example, Tour Operators Initiative Program Officers*. Paris: UNEP.

Çelik Uğuz, S. and Kaya, İ. (2016) Informal Employment in Tourism. In: Avcikurt, C. Dinu, M. Hacioğlu, N. Efe, R. Soykan, A. and Tetik, N. (eds) *Global Issues and Trends in Tourism*. Sofia: St. Kliment Ohridski University Press.

Centre of Study for Policies and Practices in Education (CEPPE) (2013) *Learning Standards, Teaching Standards and Standards for School Principals: A Comparative Study*. OECD Education Working Papers, No. 99. Chile: OECD Pub.

Chen, L. and Chen, J. (2011) The Motivations and Expectations of International Volunteer Tourists: A Case Study of Chinese Village Traditions. *Tourism Management* 32, 435–442.

Christiansen, D. (1990) Adventure Tourism. In: Miles, J. and Priest, S. (eds) *Adventure Education*. State College, PA: Venture Publishing.

Chron (2018) How to Improve Call Centre English. Available at: http://work.chron.com/improve-call-center-english-13785.html (accessed 22 June 2018).

Clark, D. (1993). Teacher Evaluation: A Review of the Literature with Implications for Educators. Seminar Paper, California State University at Long Beach.

Cloutier, R. (2003) The Business of Adventure Tourism. In: Hudson, S. (ed.) (2003) *Sport and Adventure Tourism*. USA: Haworth Hospitality Press.

Coard, B. (2005) Why I wrote the 'ESN book'. *Guardian*, 5 February 2005. Available at: http://www.theguardian.com/education/2005/feb/05/schools.uk (accessed 15 March 2018).

Coghlan, A. (2006) Choosing your Conservation-Based Volunteer Tourism Market Segment with Care – Part1. *Voluntourist*. Available at: http://www.voluntourism.org/news-studyandresearch0106.htm (accessed 3 December 2017).

Coghlan, A. and Fennell, D. (2009) Myth or Substance: An Examination of Altruism as the Basis of Volunteer Tourism. *Annals of Leisure Research* 12, 377–403.

Coghlan, A. and Gooch, M. (2011) Applying a Transformative Learning Framework to Volunteer Tourism. *Journal of Sustainable Tourism* 19, 713–728.

Cohen, E. (1972) Towards a Sociology of International Tourism. *Social Research* 39, 164–182.

Cohen, E. (1973) Nomads from Affluence: Notes on the Phenomenon of Drifter Tourism. *International Journal of Comparative Sociology* 14, 89–103.

Cohen, E. (1974) Who is a Tourist? A Conceptual Clarification. *Sociology* 22, 527–555.

Cohen, S. (2014) Consumer Behaviour in Tourism: Concepts, Influences and Opportunities. *Current Issues in Tourism* 17, 872–909.

Coleman, H. (2011) *Dreams and Realities: Developing Countries and the English Language*. London: British Council.

Collins, M. (1990) *Marva Collins' Way: Returning to Excellence in Education*, 2nd edn. New York: Jeremy P. Tarcher

Complete University (2017) Distance Learning. Available at: https://www.thecompleteuniversityguide.co.uk/distance-learning/ (accessed 27 October 2018).

Coren, N. and Gray, T. (2012) Commodification of Volunteer Tourism: A Comparative Study of Volunteer Tourists in Vietnam and in Thailand. *International Journal of Tourism Research* 14, 222–234

Council of Europe (2018) The CEFR Levels. Available at: https://www.coe.int/en/web/common-european-framework-reference-languages/level-descriptions (Accessed 28 June 2018).

Crompton, J. (1979) Motivations for Pleasure Vacations. *Annals of Tourism Research* 6, 408–424.

Csapo, J. (2012) The Role and Importance of Cultural Tourism in Modern Tourism Industry. In: Kasimoglu, M. (ed.) *Strategies for Tourism Industry – Micro and Macro Perspectives*. Rijka, Croatia: Intech.

Daldeniz, B. and Hampton, M. (2011) VOLUNtourists versus volunTOURISTS: A True Dichotomy or Merely a Differing Perception? In: Benson, A. (ed.) *Volunteer Tourism: Theoretical Frameworks and Practical Applications*. London: Routledge.

Dangerous Business (2017) The Traveller vs Tourist Debate and Why I Don't Give a Crap. Available at: http://www.dangerous-business.com/2011/12/the-traveler-vs-tourist-debate-and-why-i-dont-give-a-crap/ (accessed 22 March 2018).

Davidson, M. (2008) Spoiled Mixture: Where Does State-led 'Positive' Gentrification End? *Urban Studies* 45, 2385–2405. https://doi.org/10.1177/0042098008097105

Davidson, M. (2009) Displacement, Space and Dwelling: Placing Gentrification Debate. *Ethics, Place and Environment* 2, 219–234.

Davidson, M. (2010) Love Thy Neighbour? Social Mixing in London's Gentrification Frontiers. *Environment and Planning A: Economy and Space* 42, 524–544. https://doi.org/10.1068/a41379

Davidson, M. and Lees, L. (2005) New-Build 'Gentrification' and London's Riverside Renaissance. *Environment and Planning A: Economy and Space* 37, 1165–1190. https://doi.org/10.1068/a3739

Davidson, M. and Lees, L. (2010) New-build Gentrification: Its Histories, Trajectories, and Critical Geographies. *Population, Space and Place* 16, 395–411. https://doi.org/10.1002/psp.584

de Oliveira, A. and Paiva, V. (2007) Vulnerability to HIV: Tourism and the Use of Alcohol and Other Drugs. *Revista de Saúde Pública* 41, 1–6.

Department for Education (2017) School-Led Training. Available at: https://getintoteaching.education.gov.uk/explore-my-options/teacher-training-routes/school-led-training (accessed 27 October 2017).

Department for Education and Skills (2004) *Department for Education and Skills: Five Year Strategy for Children and Learners*. Available at: http://www.educationengland.org.uk/documents/pdfs/2004-five-year-strategy.pdf (accessed 11 March 2019).

Desforges, L. (1998) 'Checking Out The Planet'. Global Representations/Local Identities and Youth Travel. *In:* Skelton, T. and Valentine, G. (eds) *Cool Places: Geographies of Youth Cultures*. London: Routledge.

Deveney, B. (2005) An Investigation into Aspects of Thai Culture and its Impact on Thai Students in an International School in Thailand. *Journal of Research in International Education* 4, 153–171.

DeVerteuil, G. (2011) Evidence of Gentrification-induced Displacement among Social Services in London and Los Angeles. *Urban Studies* 48, 1563–1580. https://doi.org/10.1177/0042098010379277

DFID (Department for International Development) (2000) Sustainable Livelihoods Guidance Sheets. Available at: http://www.livelihoods.org/info/info_guidancesheets.html (accessed 21 May 2018).

DFID (Department for International Development) (2001) Pro-poor Tourism Initiatives in Developing Countries: Analysis of Secondary Case Studies. Working paper 1. Edmonton: DFID.

DFID (Department for International Development) (2017) Available at: https://www.gov.uk/government/organisations/department-for-international-development

Djité, P. (2008) *The Sociolinguistics of Development in Africa*. Clevedon: Multilingual Matters

Doxey, G. (1975) A Causation Theory of Visitor-resident Irritants: Methodology and Research Inferences. In: *Travel and Tourism Research Associations Sixth Annual Conference Proceedings*, 195–98. San Diego, September.

*Economist* (2011) *English in China*. Available at: https://www.economist.com/blogs/johnson/2011/09/english-china (accessed 22 June 2018).

Edsurge (2018) Available at: https://www.edsurge.com/product-reviews/socrative (acessed 28 November 2018).

Edwards, M. (2009) Why 'Philanthrocapitalism' is not the Answer: Private Initiatives and International Development. In: Lieshout, M. and Went, R. (eds) *Doing Good or Doing Better: Development Policies in a Globalising World*. The Hague: Scientific Council for Government Policy WRR.

Ellis, C. (2003) Participatory Environmental Research in Tourism – A Global View. *Tourism Recreation Research* 28, 45–55.

English First (2016) ESL Jobs in China. Available at: http://www.englishfirst.com/ESL-Jobs/esl-jobs-in-china/ (accessed 22 November).

Enjoy TEFL (2018) *Elephant Camp Hill Tribe Teaching*. Available at: http://www.enjoytefl.com/tefl-adventure/elephant-camp-hill-tribe-teaching/ (accessed 11 July 2018).

Erling, E.J. and Seargeant, P. (eds) (2012) *English and International Development*. Bristol: Multilingual Matters.

ESL Café (2016) *Teaching in Korea*. Available at: http://forums.eslcafe.com/korea/viewtopic.php?t=75211 (accessed 29 December 2017).

Ewert, A. and Hollenhorst, S. (1989) Testing the Adventure Model: Empirical Support for a Model of Risk. Recreation Participation. *Journal of Leisure Research* 21, 124–139.

Ewert, A. and Jamieson, L. (2003) Current Status and Future Directions in the Adventure Tourism Industry. *In:* Wilks, J. and Page, S. (eds) *Managing Tourist Health and Safety in the New Millennium*, pp. 67–84. Oxford: Pergamon.

Ewert, A. and Shultis, J. (1997) Resource-based Tourism: An Emerging Trend in Tourism Experiences. *Park and Recreation* 32, 94–104.

Fairclough, N. (1989) *Language and Power*. London: Longman.
Falk, J. and Dierking, L. (2010) The 95% Solution: School is not Where Most Americans Learn Most of their Science. *American Scientist* 98, 486–493.
Falk, J., Storksdierck, M. and Dierking, L. (2007) Investigating Public Science Interest and Understanding: Evidence for the Importance of Free-Choice Learning. *Public Understanding of Science* 16, 455–469.
Falk, J., Ballantyne, R., Packer, J. and Beckendoriff, P. (2012) Travel and Learning: A Neglected Research Area. *Annals of Tourism Research* 39, 908–927.
Fletcher, J., Fyall, A., Gilbert, D. and Wanhil, S. (2013) *Tourism: Principles and Practice*. London: Pearson.
Floric, M. and Broyles, M. (2012) *Sexual Abuse*. New York, NY: The Rosen Publishing Group, Inc.
*Forbes* (2017) Available at: https://www.forbes.com/sites/wadeshepard/2017/02/14/how-amazons-wooing-of-chinese-sellers-is-hurting-american-innovation/#72fb46a71df2 (accessed 22 August 2018).
Foucault, M. (1991) Governmentality. In: Burchell, G., Gordon, C. and Miller, P. (eds) *The Foucault Effect: Studies in Governmentality*, 87–104. Chicago: University of Chicago Press.
Fowler, R. (1975) *Language and Control*. London: Routledge.
Freeman, R. (1984) *Strategic Management: A Stakeholder Approach*. Pitman: Boston.
Frey, N. and George, R. (2010) Responsible Tourism Management: The Missing Link Between Business Owners' Attitudes and Behaviour in the Cape Town Tourism Industry. *Tourism Management* 31, 621–628.
Frontiers (2017) Ghana Teaching Project. Available at: https://www.frontier.ac.uk/projects/128/Ghana-Teaching-Project (accessed 22 February 2017).
Gap 360 (2016) Paid Teaching in Thailand. Available at: https://www.gap360.com/paid-teaching-in-thailand (accessed 29 November 2018).
GapYear (2018) *Teaching in China*. https://www.gapyear.com/products/270593/china-teaching/details (accessed 22 March 2018).
Gaventa, J. (2003) *Power After Lukes: A Review of the Literature*. Brighton: Institute of Development Studies.
Gecko, Bradt Travel Guides and Lasso Communications (2009) *Volunteer Travel Insights Report*. USA.
GoAbroad (2016) Get TEFL Certified and get Paid to Teach English in Thailand. Available at http://www.goabroad.com/providers/international-tefl-academy/programs/get-paid-to-teach-english-in-a-tropical-paradise-of-thailand-with-a-tefl-tesol-certification-85303 (accessed 8 May 2016).
Goldberg, A. (2002) Social Change Philanthropy and How it's Done. *Foundation News and Commentary* 43, 17–20.
Goldstein, H. (2001) Using Pupil Performance Data for Judging Schools and Teachers: Scope and Limitations. *British Educational Research Journal* 27, 433–443.
Goodwin, H. (2009) Reflections on 10 Years of Pro-Poor Tourism. *Journal of Policy Research in Tourism, Leisure and Events* 1, 90–94.
Goodwin, H., McCombes, L. and Eckardt, C. (2009) *Advances in Travel Philanthropy: Raising Money Through the Travel and Tourism Industry for Charitable Purposes*. London: WTM Report.
Grabowski, S. (2013) Volunteer Tourists: Why Do They Do It? In: Wearings, S. and McGehee, N. (eds) *International Volunteer Tourism: Integrating Travellers and Communities*. Wallingford: CAB International.
Graburn, H. (1989) Tourism: The Sacred Journey. In: Smith, V. (ed.). *Hosts and Guests: The Anthropology of Tourism*, 2nd edn. Philadelphia, PA: University of Pennsylvania Press, 21–36.
Gray, N. and Campbell, L. (2007) A Decommodified Experience? Exploring Aesthetic, Economic and Ethical Values for Volunteer Tourism in Costa Rica. *Journal of Sustainable Tourism* 15, 463–482.
Gray, N. and Campbell, L. (2009) A Decommodified Experience? Exploring Aesthetic, Economic and Ethical Values for Volunteer Ecotourism in Costa Rica. *Journal of Sustainable Tourism* 5, 463–482.
Gregory, R. (1970) *The Intelligent Eye*. London: Weidenfeld and Nicolson.
Griffith, S. (2014) *Teaching English Abroad: Talk your Way Around the World 13e*. Oxford:Vacation Work.
Groupon (2017a) TEFL Cambridge. Available at: https://www.groupon.co.uk/deals/tefl-cambridge-6?keyw=online%20tefl%20course&crea=221974833770&netw=g&adpos=1t4&utm_campaign=UK_DT_SEA_GGL_TXT_TTT_SR_CBP_CH1_NBR_k*online%20tefl%20course_m*e_c*221974833770_d*UK-RTC-KWService-National_g*RTC-KWService-tefl-cambridge-6-EXACT_s*1407_t*kwd- (accessed 22 May 2018).
*Guardian* (2016) Volunteer Travel: Experts Raise Concerns Over Unregulated Industry. Available at: https://www.theguardian.com/global-development-professionals-network/2016/jan/13/concerns-unregulated-volunteer-tourism-industry (accessed 11 June 2018).
*Guardian* online (2018a) English Language Teachers – UAE. Available at: https://jobs.theguardian.com/job/6586075/english-language-teachers-uae/?LinkSource=PremiumListing (accessed 22 January 2018).

*Guardian* online (2018b) Available at: https://www.theguardian.com/global-development-professionals-network/2016/apr/20/expat-wages-900-per-cent-higher-than-local-employees-study (accessed 28 November 2018)

Gumtree (2018) English Lessons. https://www.gumtree.com/p/english-classes/english-lessons-on skype one-to-one-pronunciation-exam-practise/1186522478 (accessed 12 February 2018).

Guttentag, D. (2009) The Possible Negative Impacts of Volunteer Tourism. *International Journal of Tourism Research* 11, 537–551.

Hall, C. and Weiler, B. (1992) Introduction: What's So Special about Special Interest Tourism? In: Hall, C. and Weiler, B. (eds) *Special Interest Tourism*. London: Belhaven Press.

Harbert, W., McConnell-Ginet, S., Miller, A. and Whitman, J. (2009) *Language and Poverty*. Clevdon: Miltilingual Matters.

Hayden, M. (2006) *Introduction to International Education: International Schools and their Communities*. California: Sage Publications.

Heath, S. (2007) Widening the Gap: Pre-university Gap Years and the 'Economy of Experience'. *British Journal of Sociology Education* 28, 89–103.

Hero, P. (2001) Giving Back the Silicon Valley Way: Emerging Patterns of a New Philanthropy. *New Directions for Philanthropic Fundraising* 32, 47–58.

Herzberg, F., Mausner, B. and Snyderman, B. (1959) *The Motivation to Work*, 2nd edn. New York: John Wiley & Sons.

Holmes, K. and Smith, K. (2012) *Managing Volunteers in Tourism*. London: Routledge.

Hoogvelt, M. (2001) *Globalization and the Postcolonial World: The New Political Economy of Development*. Baltimore, MD: Johns Hopkins University Press.

Horner, S. and Swarbrooke, J. (2007) *Consumer Behaviour in Tourism*. Abingdon: Routledge.

Hudson, S. (ed.) (2003) *Sport and Adventure Tourism*. USA: Haworth Hospitality Press.

Hustinx, L. (2001) Individualisation and New Styles of Youth Volunteering: An Empirical Investigation. *Voluntary Action* 3, 57–76.

Hutchinson, G.E. (1957). Concluding Remarks. *Cold Spring Harbour Symposium on Quantitative Biology* 22, 415–427.

i-to-i (2017) Teach English in Peru. Available at: https://www.i-to-i.com/teach-english-abroad/peru.html (accessed 19 January 2018).

ILO (2012) *International Standard Classification of Occupations, ISCO-08*. Geneva: ILO.

International TEFL Academy (2018) Available at: https://www.internationalteflacademy.com (accessed at 12 March 2019).

Jampian, S. (2012) Local Thai Pedagogical Practices and the Respect for International Human Right: A Case Study of Failed Corporal Punishment Ban in Thai Schools. *IPEDR*. 54(41) 99–204.

Jones, A. (2005) Assessing International Youth Service Programmes in Two Low Income Countries. *Voluntary Action: The Journal of the Institute for Volunteering Research* 7, 87–100.

Keegan, W., Moriarty, S. and Duncan, T. (1992) *Marketing*. Englewood-Cliffs, NJ: Prentice Hall.

Keese, J. (2011) The Geography of Volunteer Tourism: Place Matters. *Tourism Geographies* 13, 257–279.

Kirkpatrick, R. (2012) English Education in Thailand: 2012. *Asian EFL Journal* 61, 24–40.

Kogar, T. (2014) We Change People's Lives: Commodification of English in Thailand. *The New English Teacher* 8, 142–161.

Kolb, D. (1984) *Experiential Learning*. Englewood Cliffs, NJ: Prentice Hall.

Kramarae, C., Muriel S. and O'Barr, W. (1985) *Language and Power*. Newbury Park, CA: Sage.

Kusy, F. (1991) *Thailand*. London: Cadogan.

Leiper, N. (1979) The Framework for Tourism: Towards a Definition of Tourism, Tourist and the Tourism Industry. *Annals of Tourism Research* 1, 390–407.

Leonard, R. and Onyx, J. (2009) Volunteer Tourism: The Interests and Motivations of Grey Nomads. *Annals of Leisure Research* 12, 315–333.

Lepp, A. (2008) Discovering Self and Discovering Others Through the Taita Discovery Centre Volunteer Tourism Programme, Kenya. In: Lyons, K. and Wearing, S. (eds) *Journeys of Discovery in Volunteer Tourism*. Wallingford: CAB International.

Lo, A. and Lee, C. (2011) Motivations and Perceived Value of Volunteer Tourists from Hong Kong. *Tourism Management*, 32, 326–334.

Lomba, L., Apóstolo, J. and Mendes, F. (2009) Drugs and Alcohol Consumption and Sexual Behaviours in Night Recreational Settings in Portugal. *Adicciones* 31, 309–326.

Long, M. (2000) Focus on form in task-based language teaching. In: Lambert, R.D. and E. Shohamy (eds). *Language policy and pedagogy: Essays in honor of A Ronald Walton*. Philadelphia: John Benjamins.

LSE blog (2018) Experiences of Teaching English Abroad. Available at: http://blogs.lse.ac.uk/careers/2017/05/04/experiences-of-teaching-english-abroad/ (accessed 4 May 2018).

Lyons, K. (2003) Ambiguities in Volunteer Tourism: A Case Study of Australians Participating in a J-1 Visitor Exchange Program. *Tourism Recreation Research* 28, 5–13.

Lyons, K. and Wearing, S. (eds) (2008) *Journeys of Discovery in Volunteer Tourism: International Case Study Perspectives*. Wallingford: CAB International.

Lyons, K. and Wearing, S. (2012) Reflections on the Ambiguous Intersections between Volunteering and Tourism. *Leisure Sciences* 34, 98–93.

MacCannell, D. (1973) Staged Authenticity: Arrangements of Social Space in Tourist Settings. *American Journal of Sociology* 79, 589–603.

MacIntosh, R.W. and Goeldner, R. (1986) *Tourism: Principles, Practices, Philosophies*. New York: Wiley & Sons.

Malam, L. (2004) Performing Masculinity on the Thai Beach Scene. *Tourism Geographies* 6, 455–71.

Maslow, A.H. (1943) A Theory of Human Motivation. *Psychological Review* 50, 370–396.

Matheison, A. and Wall, G. (1982) *Tourism: Economic Physical and Social Impacts*. Harlow: Longman House.

Matthews, A. (2008) *Negotiated Selves: Exploring the Impact of Local-Global Interactions on Young Volunteer Travellers*. In: Lyons, K. and Wearing, S. (eds) *Journeys of Discovery in Volunteer Tourism*. Wallingford: CAB International

Mayr, A. (2008) *Language and Power: An Introduction to Institutional Discourse*. London: Continuum.

McAlister, D. and Ferrell, L. (2002) The Role of Strategic Philanthropy in Marketing Strategy. *European Journal of Marketing* 36, 689–705.

McGehee, N. and Andereck, K. (2008) 'Pettin' the Critters': Exploring the Complex Relationship Between Volunteers and the Voluntoured in McDowell County, West Virginia, USA, and Tijuana, Mexico. In: Lyons, K. and Wearing, S. (eds) *Journeys of Discovery in Volunteer Tourism: International Case Study Perspectives*. Wallingford: CAB International.

McGehee, N., Clemmons, D. and Lee, Y. (2009) *2008 Voluntourism Survey Report*. San Diego: VolunTo

McIntosh, A. and Zahra, A. (2005) Alternative Cultural Experiences Through Volunteer Tourism. Paper presented at the Associated for Tourism and Leisure Education (A I LAS) Catalonia, Spain.

McKercher, B. and Du Cros, H. (2002) *Cultural Tourism: The Partnership Between Tourism and Cultural Heritage Management*. New York: Haworth Press.

Methanonpphakhun, S. and Deocampo, M. (2016) Being an English Language Teacher: A Narrative Analysis of Foreign Teachers in Thailand. *The New English Teacher* 10, 1–19.

Mezirow, J. (1991) *Transformative Dimensions of Adult Learning*. San Francisco: Jossey-Bass.

Millington, K. (2005). Gap year travel - International. *Travel & Tourism Analyst* 12, 1–50.

Mitchell, C. and Weber, S. (1998) The Usable Past: Teachers (Re)Playing School. *Changing English* 5, 45–56.

Mostafanezhad, M. (2013) The Politics of Aesthetics in Volunteer Tourism. *Annals of Tourism Research* 43, 150–169.

Munsters, W. (1994) *Cultuurtoerisme*. Garant: Apeldoorn.

Mustonen, P. (2005) Volunteer Tourism: Postmodern Pilgrimage? *Journal of Tourism and Cultural Change* 3, 160–177.

Newman, K. and Wyly, E. (2006) The Right to Stay Put, Revisited: Gentrification and Resistance to Displacement in New York City. *Urban Studies* 43, 23–57.

Novelli, M. (ed.) (2005) *Niche Tourism: Contemporary Issues, Trends and Cases*. Oxford: Butterworth-Heinemann.

Novelli, M., Morgan, N., Mitchell, G. and Ivanov, K. (2016) Travel Philanthropy and Sustainable Development: The Case of the Plymouth Banjul Challenge. *Journal of Sustainable Tourism* 24, 824–845.

Nyaupane, G., Morais, D. and Dowler, L. (2006) The Role of Community Involvement and Number/Type of Visitors on Tourism Impacts: A Controlled Comparison of Annapurna, Nepal and Northwest Yunnan, China. *Tourism Management* 27, 1373–1385.

ODI (Oversea Development Institute) (2007) Can Tourism offer Pro-poor Pathways to Prosperity? Examining Evidence on the Impact of Tourism on Poverty Briefing paper. Available at: http://www.odi.org.uk/sites/odi.org.uk/files/odi-assets/publications-opi... (accessed 17 April 2018).

OECD (2001) Available at: http://www.oecd.org/site/worldforum/33703702.pdf (accessed 12 March 2019).

OECD (2009) Teaching Practices, Teachers' Beliefs and Attitudes. In: Creating Effective Teaching and Learning Environments: First Results from Talis. Available at: http://www.oecd.org/education/school/43023606.pdf (accessed 5 March 2019).

OECD (2015) OECD Employment Outlook, 251, available at http://www.oecd.org/employmentoutlook (accessed 28 Novermber 2018).

Oglivie, F. (1934) Tourist Traffic. In: Seligman, E.R.A. and Johnson, A. (eds) *Encyclopaedia of the Social Sciences*. New York: Macmillan, 661.

One Caribbean (2017) Developing a Niche Tourism Market Database for the Caribbean. Available at: http://www.onecaribbean.org/wp-content/uploads/NicheMarketsDatabase.pdf (accessed 20 March 2018).

One Step 4Ward (2018) Available at: https://onestep4ward.com (accessed 28 November 2018).

Oppermann, M. (1993) Tourism Space in Developing Countries. *Annals of Tourism Research* 20, 535–556.

O'Reilly, C. (2006) From Drifter to Gap Year Tourist: Mainstreaming Backpacker Travel. *Annals of Tourism Research* 33, 998–1017.

O'Sullivan, E. (2002) The Project and Vision of Transformative Education: Integral Transformative Learning. In: O'Sullivan, E. Morrell, A. and O'Connor, M. (eds) *Expanding the Boundaries of Transformative Learning: Essays on Theory and Praxis*. New York: Palgrave.

Overseas Development Institute (2001) Pro-Poor Tourism Strategies: Making Tourism Work For The Poor. A review of Experience. Available at: http://www.odi.org/sites/odi.org.uk/files/odi-assets/publications-opinion-files/3246.pdf (accessed 2 January 2018).

Overseas Development Institute and Associates for International Resources and Development (AIRD) (1999) *DAC Scoping Study of Donor Poverty Reduction Policies and Practices. DAC Informal Network on Poverty Reduction*. London: ODI.

Oxfam (2018) Available at: https://www.oxfam.org.uk/what-we-do/issues-we-work-on/health-and-education (accessed 13 September 2018).

Page, S., Bentley, T. and Walker, L. (2005) Scoping the Nature and Extent of Adventure Tourism Operations in Scotland: How Safe are They? *Tourism Management* 26, 381–397.

Palacios, C. (2010) Volunteer Tourism, Development and Education in a Postcolonial World:Conceiving Global Connections Beyond Aid. *Journal of Sustainable Tourism* 18, 861–878.

Papanastasiou, E. (1999). Teacher evaluation. Unpublished manuscript, Michigan State University, East Lansing.

Paramo La Haine, D. (2008) Tourism – Consumerism. www.ecotonline.org/publications-/resources-/doc_download/173-tourism-consumerism (accessed 20 March 2018).

Park, J. and Wee, L. (2012) *Linguistic Capital and Language Policy in a Globalizing World*. London: Routledge.

Pearce, P. and Foster, F. (2007) A University of Travel: Backpacker Learning. *Tourism Management* 28, 1285–1298.

Pells, R. (2015) Why are so Many Teachers Quitting the Classroom? *The Independent*. Available at: https://www.independent.co.uk/news/education/schools/why-are-so-many-teachers-quitting-the-classroom-10179513.html (accessed 30 November 2018).

PepsiCo (2016) Philanthropy. Available at: http://www.pepsico.com/sustainability/philanthropy (accessed 2 May 2019).

Pereira, P. (2005) Becoming a Teacher of Mathematics. *Studying Teacher Education* 1, 69–83.

Petroman, I., Petroman, C., Marin, D., Ciolac., Vaduva, L. and Pandur, I. (2013) Types of Cultural Tourism. *Sientific Papers: Animal Science and Biotechnologies* 46, 385–388.

Plog, S. (1987) Understanding Psychographics in Tourism Research. In: Ritchie, J. and Goeldner, C. (eds) *Travel, Tourism, and Hospitality Research*. New York: John Wiley.

Political Studies Association (2018) Available at: https://www.psa.ac.uk/sites/default/files/TEF%20Bulletin.pdf (accessed 28 November 2018).

Poon, A. (1993) *Tourism Technology and Competetive Strategies*. Wallingford: CAB International.

Porter, M. and Kramer, M. (2002) The Competitive Advantage of Corporate Philanthropy. *Harvard Business Review* 80, 56–68.

Projects Abroad (2018) Available at: https://www.projects-abroad.co.uk/about-us/money-matters/ (accessed 28 November 2018).

Prospects (2017) Responsibilities. Available at: https://www.prospects.ac.uk/job-profiles/english-as-a-foreign-language-teacher (accessed 11 January 2018).

Pumwaree, S. (1986) *Effects of Differential Reinforcement of Low Rates (DRL) on Reducing Disruptive Behaviors of Mathayom Suksa One Students*. Bangkok: Graduate School, Department of Psychology, Chulalongkorn University.

Punthumasen, P. (2007) International Program for Teacher Education: Approach to Tackling Problems of English Education in Thailand. *The 11th UNESCO-APEID International Conference Reinventing Higher Education: Toward Participatory and Sustainable Development*. 12–14 December. Bangkok.

Purvis, K. and Kennedy, L. (2016) Volunteer Travel: Experts Raise Concerns over Unregulated Industry. *Guardian*, 13 January 2016. Available at: https://www.theguardian.com/global-development-professionals-network/2016/jan/13/concerns-unregulated-volunteer-tourism-industry (accessed 31 January 2019).

Putnam, R. (2000) *Bowling Alone: The Collapse and Revival of American Community*. New York: Simon & Schuster.

Rabinow, P. (1991) *The Foulcault Reader: An Introduction to Foulcault's Thought*. London: Penguin.

Raymond, E. and Hall, C. (2008) The Development of Cross-Cultural (Mis)Understanding Through Volunteer Tourism. *Journal of Sustainable Tourism* 16, 530–543.

Read, S. (1980) A Prime Force in the Expansion of Tourism in the Next Decade: Special Interest Travel. In: Hawkins, D. Shafor, E. and Rovelstad, J. (eds) *Tourism Marketing and Management Issues*. Washington, DC: George Washington Press, 193–202.

Reddy, R. (1980) Individual Philanthropy and Giving Behaviour. In: Horton-Smith, D. Macaulay, J. *et al.* (eds) *Participation in Social Political Activities*. San Francisco: Jossey-Bass.

Redmon, D. (2003) Playful Deviance as an Urban Leisure Activity: Secret Selves, Self-Validation, and Entertaining Performances. *Deviant Behavior* 24, 27–51.

Reisinger, Y. (2009) *International Tourism Cultures and Behaviour*. Oxford: Elsevier.

Richards, G. (2005) *Cultural Tourism in Europe*. Wallingford: CAB International.

Richter, L. and Norman, A. (2010) AIDS Orphan Tourism: A Threat to Young Children in Residential Care. *Vulnerable Children and Youth Studies: An International Interdisciplinary Journal for Research, Policy and Care* 5, 217–229.

Ritchie, B., Carr, N. and Cooper, C. (2003) *Managing Educational Tourism*. Clevedon: Channel View.

Ritzer, G. (1996). *The McDonaldization of Society : An Investigation into the Changing Character of Contemporary Social Life*. Thousand Oaks, CA: Pine Forge Press.

Robertson, R. (1992) *Globalization: Social Theory and Global Culture*. London: Sage.

Robinson, M. and Novelli, M. (2007) Niche Tourism: An Introduction. In: Novelli, M. (ed.) *Niche Tourism*. London: Routledge.

Rojek, C. (1983) Emancipation and Demoralisation: contradicting approaches in the sociology of leisure Leisure Studies, 2, 83–96.

Rostow W. (1960) *The Stages of Growth: An Anti Communist Manifesto*. Cambridge: Cambridge University Press.

Royal Thai Embassy (2016) Visa for English Teacher. http://www.thaiembassy.org/en/services/29705-Visa-for-English-Teacher.html (accessed 15 February 2016).

Ruenkaew, P. (1999) Marriage Migration of Thai women to Germany. International Institute of Asian Studies and the University of Amsterdam. *7th International Conference on Thai studies*, Amsterdam.

Sailor, D. (2010) *Supporting Children in their Home, School and Community*. New Jersey: Prentice Hall.

Sanders, W. and Rivers, J. (1996) *Cumulative and Residual Effects of Teachers on Future Student Academic Achievement*. Knoxville: University of Tennessee Value-Added Research and Assessment Center.

Scam.com (2018) *China Scam Patrol*. Available at: https://www.scam.com/showthread.php?704467-UPDATED-China-Liars-List-ESL-TEFL-Teacher-Job-Scams-Internships-Exporters-etc-BLACKLIST&p=1970246#post1970246 (accessed 22 January 2018).

Scarini, J. and Pearce, P. (2012) The Perceived Influence of Travel Experiences on Learning Generic Skills. *Tourism Management* 33, 380–386.

SEE TEFL (2018) Available at: https://seetefl.com/reasons-to-choose-see-tefl-for-teacher-training/ (accessed 28 November 2018).

Sharpley, R. (2006) *Travel and Tourism*. London: Sage.

Sharpley, R. and Telfer, D. (2002) *Tourism and Development: Concepts and Issues*. Clevedon: Channel View.

Silk, J. (2004) Caring at a Distance: Gift Theory, Aid Chains and Social Movements. *Social & Cultural Geography* 5, 229–251.

Simpson, K. (2004) 'Doing Development': The Gap Year, Volunteer-tourists and a Popular Practice of Development. *Journal of International Development* 16, 681–692.

Sims, J. (2012) Beyond the Stereotype of the 'Thai-Bride'. Visibility, Invisibility and Community. In: Charsley, K. (ed.) *Transnational Marriage: New Perspectives from Europe and Beyond*. London: Routledge.

Sin, H. (2009) Volunteer Tourism: 'Involve me and I will Learn'? *Annals of Tourism Research* 36, 480–501.

Skinner, B.F. (1948) 'Superstition' in the Pigeon. *Journal of Experimental Psychology* 38, 168.

Slater, T. (2009) Missing Marcuse: On Gentrification and Displacement. *City: Analysis of Urban Trends, Culture, Theory, Policy, Action* 13, 292–311.

Soderman, N. and Snead, S. (2008) Opening the Gap: The Motivation of Gap Year Travellers to Volunteer in Latin America. In: Lyons, K. and Wearing, S. (eds) *Journeys of Discovery in Volunteer Tourism: International Case Study Perspectives*. Wallingford: CAB International.
Stainton, H. (2016) A Segmented Volunteer Tourism Industry. *Annals of Tourism Research* 16, 250–250.
Stainton, H. (2017a) Inside the Thai Classroom: A Stakeholder Insight. *The New English Teacher* 11, 1–20.
Stainton, H. (2017b) The Commodification of English Language Teaching in Tourism: A Sustainable Solution? *Tourism Management Perspectives* 25, 123–130.
Stainton, H. (2017c) TEFL Tourism: A Phenomenological Examination of the TEFL Teacher in Thailand. PhD thesis, Buckingham New University.
Stainton, H. (2017d) Available at: https://www.tes.com/jobs/careers-advice/teaching-overseas/tefl-overseas-uk-classroom (accessed 28 November 2018).
Stainton, H. (2018a) The Commodification of English Language Teaching in Tourism: A Sustainable Solution? *Tourism Management Perspectives* 25, 123–130.
Stainton, H. (2018b) TEFL Tourism: The Tourist Who Teaches. *Tourism Geographies* 20, 127–143. https://doi.org/10.1080/14616688.2017.1298151
Stebbins, R.A. (1992). *Amateurs, Professionals and Serious Leisure*. Montreal, QC: McGill-Queen's University Press.
Stebbins, R.A. (1997) Casual Leisure: A Conceptual Statement. *Leisure Studies* 16, 17–25.
Stebbins, R.A. (2000) Obligation as an Aspect of Leisure Experience. *Journal of Leisure Research* 32, 152–155.
Stebbins, R.A. (2005) Project-based Leisure: Theoretical Neglect of a Common use of Free Time. *Leisure Studies* 24, 1–11.
Stebbins, R.A. (2007) *Serious Leisure: A Perspective for our Time*. New Brunswick, NJ: Transaction.
Stoddart, H. and Rogerson, C. (2004) Volunteer Tourism: The Case of Habitat for Humanity South Africa. *Geojournal* 60, 311–318.
Stone, N. and Petrick, J. (2013) The Educational Benefits of Travel Experiences: A Literature Review. *Journal of Travel Research* 52, 731–744.
Street, B. (2001) *Literacy and Development: Ethnographic Perspectives*. London: Routledge.
Swank, P.R., Taylor, R.D., Brady, M.P. and Freiberg, H.J. (1989) Sensitivity of Classroom Observation Systems: Measuring Teacher Effectiveness. *Journal of Experimental Education* 57, 171–186.
Swarbrooke, J. (1999) *Sustainable Tourism Management*. Wallingford: CAB International.
Swarbrooke, J., Beard, C., Leckie, S. and Pomfret, G. (2003) *Adventure Tourism: The New Frontier*. London: Butterworth-Heinemann.
Taillon, J. and Jamal, T. (2013) Understanding the Volunteer Tourist: A Qualitative Enquiry. *VolunTourist* 4. Available at: http://www.voluntourism.org/news-studyandresearch44.htm (accessed 8 December 2017).
TALIS (2014) Available at: http://www.oecd.org/education/school/UK-TALIS-2013-national-report.pdf (accessed 12 December 2018).
Teach Away (2016) Teaching Jobs in Thailand. Available at: https://www.teachaway.com/teach-thailand (accessed 30 November 2018).
Teach English ESL (2016) Krabi TEFL Course in Thailand. http://teachenglishesl.com/tefl-teaching-certificate/krabi-tefl-course/(accessed 8 May 2016).
TEFL Heaven (2017) Teach English in Costa Rica. https://www.teflheaven.com/teaching-abroad-programs/teach-english-in-costa-rica-tefl-heaven-manuel-antonio-paid-teaching-job-costa-rica/ (accessed 12 March 2019).
TESOL International Association (2014) Available at: https://www.tesol.org/ (accessed 12 March 2019).
The School Fund (2018) Available at: https://theschoolfund.org/cgi-bin/dyn?c=info&t=about&m=story (accessed 22 August 2018).
The Serious Leisure Perspective (2013) Diagrams of the Serious Leisure Perspective. Available at: http://www.seriousleisure.net/slp-diagrams.html (accessed 5 January 2018).
This is Money (2017) Available at: http://investing.thisismoney.co.uk/quote/TUI (accessed 22 June 2018).
TIC Recruitment (2018) Available at: https://www.ticrecruitment.com/blog/why-are-teachers-in-the-uk-leaving-the-profession (acessed 28 November 2018).
Tipps C. (1973) *Modernization Theory and the Comparative Study of Societies: A Critical Perspective*. Cambridge: Cambridge University Press.
TOI (Tour Operators Initiative) (2005) Integrating Sustainability into Business. Management Guide for Responsible Tour Operators. http://www.toinitiative.org/fileadmin/docs/publications/Sustainability_in_Business_-_Management.pdf (accessed 12 May 2018).

TOI (Tour Operators Initiative) (2007) Good practice. Http://www.toinitiative.org/index.php?id=48. (accessed 18 May 2018).

Tomazos, K. and Butler, R. (2009) Volunteer Tourism: Working on Holiday or Playing at Work? *Tourismos: An International Multidisciplinary Journal of Tourism* 4, 331–349.

Tomazos, K. and Butler, R. (2010) The Volunteer Tourist as 'Hero'. *Current Issues in Tourism* 13, 363–380.

Tomazos, K. and Cooper, W. (2012) Volunteer Tourism: At the Crossroads of Commercialisation and Service? *Current Issues in Tourism* 15(5) 405–423.

Tourism Concern (2014) Available at: https://tourismconcern.org.uk/wp-content/uploads/2014/06/Annual-Report-2014.pdf (accessed 12 March 2019).

Tourism Research and Marketing (TRAM) (2008) *Volunteer Tourism: A Global Analysis*. Arnhem: ATLAS Publications.

Troyna, B. and Williams, J. (1986) *Racism, Education and the State*. London: Taylor & Francis.

Tsui, B. (2000) Marketing Adventures. *Advertising Age* 71, 38.

TUI (2018) Available at: https://www.tui.co.uk/ (accessed 12 March 2019).

Turner, B. (1994) *Orientalism, Post-modernism and Globalism*. London: Routledge.

Tutenges, S. (2013) Stirring up Effervescence: An Ethnographic Study of Youth at a Nightlife Resort. *Leisure Studies* 32, 233–248.

UNESCO (2005) Available at: https://en.unesco.org/gem-report/report/2006/literacy-life (accessed 12 December 2018).

UNESCO (2012) *Why Language Matters for the Millenium Development Goals*. Bangkok: UNESCO.

UNESCO (2016) Teaching and Learning: Achieving Quality for All. Available at: http://www.unesco.org/new/fileadmin/MULTIMEDIA/HQ/ED/GMR/pdf/Malawi_Factsheet.pdf (accessed 22 July 2018).

UNESCO (2018) Available at: http://www.unesco.org/new/en/social-and-human-sciences/themes/international-migration/glossary/poverty/ (accessed 29 November 2018).

UNESCO Centre for Global Development (2013) Available at: https://unesdoc.unesco.org/ark:/48223/pf0000226992 (accessed 2 May 2019).

UNICEF (2012) The State of the World's Children. Available at: https://www.unicef.org/sowc/index_61804.html (accessed 12 March 2018).

United Nations (1987) *Our Common Future - Brundtland Report*. Oxford: Oxford University Press.

United Nations (2011) Gross National Income Converted to International Dollars Using Purchasing Power Parity Rates. Available at: http://www.un.org/en/development/desa/news/2011.html (accessed 28 November 2018).

United Nations (2018) Millennium Development Goals. Available at: http://www.un.org/millenniumgoals/education.shtml (accessed 28 November 2018).

University of Canberra (2009) *Gap Year Sets Future Students on Course for UC*. www.canberra.edu.au/gap-year-plus/home (accessed 10 June 2018).

UNWTO (2014) UNWTO Tourism Highlights 2014 edition. Available at: https://www.e-unwto.org/doi/pdf/10.18111/9789284416226) (accessed 1 February 2019).

Uriely, N. (2001) 'Travelling Workers' and 'Working Tourists': Variations Across the Interaction Between Work and Tourism. *International Journal of Tourism Research* 3, 1–8.

Uriely, N. and Belhassen, Y. (2005) Drugs and Tourists' Experiences. *Journal of Travel Research* 43, 238–246.

Urry, J. (1990) *The Tourist Gaze: Leisure and Travel in Contemporary Societies*. London: Sage.

Vainikka, V. (2014) Travel Agent Discourses of Mass Tourism: Beyond Stereotypes? *Tourism Geographies* 16, 318–332.

Valdez, A. and Sifaneck, S. (1997) Drug Tourists and Drug Policy on the US-Mexican Border: An Ethnographic Investigation. *Journal of Drug Issues* 27, 879–898.

van't Klooster, E., van Wijk, J., Go, F. and van Rekom, J. (2008) Educational Travel. The Overseas Internship. *Annals of Tourism Research* 35, 690–711.

Vermeylen, G. (2008). Informal Employment in the European Union. Available at: http://wiego.org/sites/wiego.org/files/publications/files/Vermeylen_European_Union.pdf (accessed 28 November 2018).

Vice (2016) Evil people are Exploiting Cambodias Orphans. Available at: https://www.vice.com/en_uk/article/xd4bp4/cambodian-orphanages (accessed 12 March 2019).

Vogt, C. (1984) Developing a Teacher Evaluation System. *Spectrum* 2, 41–46.

Volunteer England (2008) Definitions of Volunteering. *Volunteering England Information Sheet*. London: Volunteering England.

VSO (2017) Language Trainer Myanmar. Available at: https://www.vsointernational.org/news/blog/5-minutes-alice-redfearn-english-language-trainer-myanmar (accessed 12 February 2018).

Wagner, L. (2002) The 'New' Donor: Creation or Evolution? *International Journal of Nonprofit and Voluntary Sector Marketing* 7, 342–353.

Walker, H., Ramsey, F. and Gresham, F. (2004) *Antisocial Behavior in School: Evidence-Based Practices,* 2nd edn. Belmont, CA: Wadsworth, Cengage Learning.

Wearing, S. (2001) *Volunteer Tourism: Experiences that make a Difference*. Wallingford: CAB International.

Wearing, S. (2002) Re-Centering the Self in Volunteer Tourism. In: Dann, G. (ed.) *The Tourist as a Metaphor of the Social World*. Wallingford: CAB International.

Wearing, S. (2004) Examining Best Practice in Volunteer Tourism. In: Stebbins, R. and Graham, M. *Volunteering as Leisure/Leisure as Volunteering: An International Assessment*. Wallingford: CAB International.

Wearing, S. and McGehee, N. (2013) *International Volunteer Tourism. Integrating Travellers and Communities*. Wallingford: CAB International.

Wearmouth, J. and Glynn, T. (2013) Issues in Inclusion and the Management of Student Behaviour in Schools. In: Wearmouth, J., Glynn, T., Richmond, R. and Berryman, M. (eds) *Inclusion and Behaviour Management in Schools: Issues and Challenges*. London: Routledge.

Weaver, D. (2007) *Sustainable Tourism*. Oxford: Elsevier Butterworth-Heinemann.

Weber, K. (2001) Outdoor Adventure Tourism: A Review of Research Approaches. *Annals of Tourism Research* 28, 363–380.

Weber, M. (1978) *Economy and Society: An Outline of Interpretive Sociology*. Oakland: University of California Press.

Wenglinsky, H. (2000) *How Teaching Matters: Bringing the Classroom Back into Discussions of Teacher Quality*. Educational Testing Service, Princeton, NJ. Policy Information Center.

Werry, M. (2008) Pedagogy of/as/and Tourism: Or, Shameful Lessons. *Review of Education, Pedagogy, and Cultural Studies* 30, 14–42.

Wickens, E. (1997) Licensed for Thrill: Risk-Taking and Tourism. In: Clift, S. and Gabowski, P. *Tourism and Health*. London: Printer.

Wickens, E. (2011) Journeys of the Self. Volunteer Tourists in Nepal. In: Benson, A. (ed.) *Volunteer Tourism. Theory Framework to Practical Applications*. London: Routledge.

Wijk, V. and Persoon, W. (2006) A Long-haul Destination: Sustainability Reporting Among Tour Operators. *European Management Journal* 24, 381–295.

Wilks, J. and Page, S. (eds) (2003) *Managing Tourist Health and Safety in the New Millennium*. Oxford: Pergamon.

Williams, A. and Soutar, G. (2005) Close to the Edge: Critical Issues for Adventure Tourism Operators. *Asia Pacific Journal of Tourism Research* 10, 247–262.

Woolard, K.A. (1998) Introduction: Language Ideology as a Field of Inquiry. In: Schieffelin, B.B. Woolard, K.A. and Kroskrity, P.V. (eds) *Language Ideologies: Practice and Theory*. Oxford: Oxford University Press.

Wright, H. (2014) Volunteer Tourism and its (Mis) Perceptions: A Comparative Analysis of Tourist/Host Perceptions. *Journal of Tourism and Hospitality Research* 13, 239–250.

WTO (World Tourism Organisation) (2000) Cultural Heritage and Tourism Development. *Report on the International Conference on Cultural Tourism*. Cambodia: WTO.

WTO (World Tourism Organisation) and European Commission (2013) *Sustainable Tourism for Development*. Available at: http://cf.cdn.unwto.org/sites/all/files/docpdf/devcoengfinal.pdf (accessed 2 May 2019).

World Trade Organisation and United Nations Statistics Division (1994) *Recommendations on Tourism Statistics*. Madrid New York: World Trade Organisation.

Yeoman, I. (2008) *Tomorrow's Tourist. Scenarios and Trends*. London: Routledge

Yeoman, I. (2012) *2050 – Tomorrow's Tourism*. Bristol: Channel View Publications.

Youth Central (2018) Gap Years. Available at: www.youthcentral.vic.gov.au (accessed 20 July 2018).

Zuckerman, M. (1994) *Behavioural Expressions and Biosocial Bases of Sensation Seeking*. Cambridge: Cambridge University Press.

# Index